MICRO-FACTS

Sixth Edition

Revised by
Peter Wareing and Rhea Fernandes

Originally compiled by
Aynsley Halligan

This edition first published 2007 by
Leatherhead Publishing
a division of
Leatherhead Food International Ltd
Randalls Road, Leatherhead, Surrey KT22 7RY, UK
URL: http://www.leatherheadfood.com

and

Royal Society of Chemistry
Thomas Graham House, Science Park, Milton Road,
Cambridge, CB4 0WF, UK
URL: http://www.rsc.org
Registered Charity No. 207890

ISBN: 978-1-905224-43-2

A catalogue record of this book is available from the British Library

Typeset by Alison Turner
Printed and bound by Cpod

FOREWORD

The previous five editions of 'Micro-Facts' have proved a useful ready reference for all those who are concerned with food safety. In this sixth edition, we have reviewed the entire book for currency of information. Where necessary, we have updated the text with new information on conditions for growth, sources of the organism concerned, and any new data on foodborne disease caused by the organism.

We have limited our update to information from the past few years, and reviewed the older references. However, we have retained older references where these are not supplanted by more recent information, or if it is a key historical reference. In this way we have updated the information, without losing the compact nature of the book. Our emphasis continues to be to serve the needs of the food industry, whether a manufacturer, retailer or caterer.

A key change in the sixth edition is the promotion of *Enterobacter sakazakii* to a full chapter, given the recent rise in prominence of this organism for the producers and users of infant formula and weaning products. Other changes include information on coliforms as indicators, and a new chapter on moulds and mycotoxins, and yeasts and moulds as spoilage organisms. In many cases, product recalls are initiated by excessive growth of spoilage organism, not food-borne pathogens.

Acknowledgements

Micro-Facts has a long history of writing and review, originally compiled and revised by Aynsley Halligan, with the help of many experts in food microbiology, from industry, research institutions, and government. Their contribution to these editions is gratefully acknowledged. In the production of this sixth edition of the book. We are indebted to Alison Carnell, Sherok Abbas-Majid and Jane Smith, who have provided valuable assistance with the updating. I would also like to thank Catherine Raw for her careful editing, Alison Turner for typesetting the manuscript and Ann Pernet for the indexing.

CONTENTS

FOODBORNE BACTERIAL PATHOGENS

INTRODUCTION

Food Poisoning – a Brief Overview

Foodborne illness is a major public health concern worldwide. The costs in terms of human illness and economic losses to individual companies and to the public health sector can be immense. Recalls, closure of factories, legal proceedings and adverse publicity to food companies involved in 'food scares' can result in both acute and long-term losses, with repercussions that may continue for many years. For example, sales of corned beef after the 1964 typhoid outbreak in Aberdeen, Scotland, were not restored to pre-outbreak levels for about twenty years. More recently, an outbreak of *E. coli* O157 infection in the UK caused the closure of a meat processor and was the subject of major headlines in the media, causing as it did the death of a five year old boy as a result of eating a contaminated product.

We can see by observing the scientific literature and reviewing the media that foodborne illness is a worldwide problem, not restricted to any particular country. Solutions to problems will often have to be considered from outside the European community if they are to be effective. Nevertheless, there was a real increase in food poisoning in the UK in the 1980s and 1990s, reaching a peak of approximately 100,000 reported cases in 1997/1998.

There were many different causes for this increase, and the 'Richmond' Committee on the Microbiological Safety of Food concluded in 1990 that poultry and their products were the most important source of human gastrointestinal infections arising from food, both from carcass meat contaminated with *Salmonella* and *Campylobacter*; and shell eggs that could also be a source of *Salmonella*. They considered that many factors were involved in this increase, with no factor more important than the next.

More recently, the number of recorded cases of food poisoning in the UK has fallen significantly, and this is thought to be due principally to action taken by the poultry industry to eliminate *Salmonella* from laying flocks and broilers by vaccination programmes and improved hygiene. In contrast, however, reports of *Campylobacter* infection have risen over the same period, and the organism is now the most important cause of bacterial foodborne disease in the UK. The same trends are noted in other European countries and the US, and much attention is now being directed to the control of *Campylobacter* in poultry. Some countries remain free of *Salmonella* and *Campylobacter*, by the effective application of biosecurity measures designed to contain the organism.

The global nature of the food chain itself can cause problems, with a lack of control over the ingredient supply chain resulting in large-scale chemical contamination issues. Several recent large food poisoning outbreaks in the UK, caused by *Salmonella*, have had, at their root, the purchase of (presumably) cheaper eggs from outside the UK, where there is no vaccination policy.

'Emerging' pathogens are the subject of much research and discussion. They continue to present new challenges for identification and control. Changing consumer demands mean that some unlikely organisms may emerge over the next 10 years, and it is the job of microbiologists to anticipate and control these new organisms. It is important to remember that less than 20 years ago, *L. monocytogenes* was an obscure species, almost unknown outside the veterinary field. *Campylobacter* has emerged more recently as a significant cause of food poisoning from the consumption of poultry, and *E. coli* O157 as a cause of food poisoning in ground beef.

An explanation of the term 'food poisoning'

For the purposes of Micro-Facts, the rather loosely applied term 'food poisoning' encompasses all those types of illness that are caused by the consumption of food contaminated with pathogenic micro-organisms and/or their toxins. In this context, therefore, *Campylobacter* has been included as a 'food-poisoning' bacterium, despite the fact that it does not grow in food - food acts only as a vehicle for the organism.

In most cases food poisoning involves gastroenteritis - vomiting and/or diarrhoea - but in the cases of botulism and listeriosis, the main symptoms are caused by effects on other ('extra-intestinal') parts of the body.

Intoxication vs infection

Food poisoning can be split further into three types: 'intoxication', 'infection', or 'intermediate'.

Infection is caused by the *in vivo* multiplication of bacteria that are taken in with food; subsequently, live food-poisoning organisms (usually many) need to be ingested and there is normally a delayed response (typically involving diarrhoea) reflecting the time needed for an infection to develop. Examples of food-poisoning bacteria that cause infection are *Salmonella*, *Campylobacter* and *Vibrio parahaemolyticus*.

Intoxication is caused by the ingestion of toxin(s) that has (have) been pre-formed in the food. Hence, there is no requirement for live organisms to be present and the onset of symptoms (usually starting with vomiting) is soon after the toxic food is ingested. Examples are *Bacillus cereus* and *Staphylococcus aureus* intoxication.

The 'intermediate' type of food poisoning is caused by the formation of toxin(s) in the bowel as a consequence of consumption of contaminated food, as in the case of *Clostridium perfringens* food poisoning. However, for the purposes of Micro-Facts, the 'intermediate type' is included under 'Infection'.

In each individual section, descriptions are provided of all the main food-poisoning bacteria, including those that cause intoxication, infection, or the intermediate type of food poisoning. Micro-organisms that are known to be occasionally associated with foods, but not to cause 'food-poisoning' symptoms (viz. the causative agents of tuberculosis, typhoid, brucellosis, Q-fever, etc.), are not included. Other foodborne - or potentially foodborne - pathogens are also mentioned, as an aid to awareness.

Summaries of the key characteristics of the main organisms considered are to be found in the convenient tables to follow.

INFECTIONS
Food-poisoning characteristics

	Incubation time*	Symptoms*	Duration*	Mortality rate	Infective dose
Salmonella	8-72 h	Abdominal pain diarrhoea, nausea, fever, (vomiting)	2-5 d	Rare but important exceptions	Usually high (10^6+), low in some foods (10-100 cells)
Campylobacter	1-11 d Usually 2-5 d	Fever, diarrhoea (sometimes with blood & mucus), nausea, abdominal pain	1-7 d+	Rare	Low, 50-500 cells
L. monocytogenes	3-70 d ‡	Flu-like symptoms, meningitis, septicaemia, meningoencephalitis	Variable	30-40%	Not known
Y. enterocolitica	1-11 d	Abdominal pain,† diarrhoea, fever, others, incl. pharyngitis	1-3 weeks	Rare	Not known probably high $>10^4$
V. parahaemolyticus	4-96 h Usually 12-24 h	Diarrhoea (in severe cases with blood & mucus), abdominal pain, nausea	1-7 d Usually 3-5 d	Rare	10^5+
Cl. perfringens	8-22 h	Diarrhoea, abdominal pain	1-2 d	Rare	High: 10^5/g+
Verocytotoxigenic *E. coli*	1-14 d Usually 3-4 d	Diarrhoea, abdominal pain	5-10 d	1%**	Very low, between 2 and 2,000 cells
E. sakazakii	18-30 d (premature babies)	Ventriculitis, brain abscess hydrocephalus, neonatal necrotising enterocolitis		40-80% in neonates	Not known

* Typical - can vary quite widely.
‡ Depends on dose, immune state of the host, and virulence of the strain.
† Can resemble appendicitis.
** VTEC-associated HUS infection.

INFECTIONS
Growth/survival characteristics

	Temperature (°C)		Heat resistance	Min pH*	Min a_w*	Aerobic/anaerobic
	Min*	Optimum				
Salmonella	4	35-37	Heat-sensitive† $D_{60\,°C}$ 1-10 min.	3.8‡	0.93	Facultative
Campylobacter	30	42-43	Slightly more heat sensitive than *Salmonella* $D_{55\,°C}$ 0.74-1 min.	4.9	0.98	Fastidious micro-aerophile – capnophilic
L. monocytogenes	-0.4	30-37	Slightly less heat sensitive than *Salmonella* $D_{60°\,C}$ 5-8 min.	4.3	0.9 in glycerol	Facultative
Y. enterocolitica	-2	28-29	Heat-sensitive $D_{60°\,C}$ 27 sec	4.2	0.95	Facultative
V. parahaemolyticus	5	30-35	Heat-sensitive $D_{41°\,C}$ 0.8-65.1 min	4.8	0.92 **	Facultative
Cl. perfringens	15	43-45	Forms heat-resistant spores	5	0.93	Anaerobic
Verocytotoxigenic *E. coli*	7	37	Heat sensitive $D_{63\,°C}$ 0.5 min	4.0 - 4.4	0.97	Facultative
E. sakazakii	5.5	37-40	Thermo-tolerant $D_{60\,°C}$ 2.5 min	3	0.2	Facultative

* Under otherwise optimal conditions - limits will vary according to strain, temperature, type of acid (in the case of pH), solute (in the case of a_w) and other factors. They will normally be higher in foods. However, variabilities in measurement, etc., must be allowed for - a margin of error must be incorporated.

† *S. Senftenberg* 775W is much more heat-resistant at high a_w. D-values for other strains of *Salmonella* increase at lower a_w, approaching or exceeding those for *S. Senftenberg* 775W.

‡ Most *Salmonella* serotypes will not grow below pH 4.5.

** Has a minimum requirement for salt – it is a halophile

INTOXICATIONS
Growth/survival characteristics

	Temperature (°C)		Heat resistance		Min pH*	Min a_w*	Aerobic/anaerobic
	Min*	Optimum	Spores	Toxins			
Cl. botulinum							
Group I	10	37	$D_{121\,°C}$ 0.21 min.	Destroyed by 5 min at 85 °C	4.6	0.94	Anaerobic
Group II	3	25-30	$D_{100\,°C}$ <0.1 min.		5	0.97	
Staph. aureus	7	30-37	NA	Heat-resistant $D_{56\,°C}$ 1-2 min in phosphate buffer	4.2†	0.86**	Facultative
B. cereus	4	30-35	$D_{95\,°C}$ 1.2-36 min.	Emetic toxin, extremely heat-resistant	4.3	0.95x	Facultative

* Under otherwise optimal conditions – limits will vary according to strain, temperature, type of acid (in the case of pH), solute (in the case of a_w) and other factors. They will normally be higher in foods. However, variabilities in measurement, etc., must be allowed for – a margin of error must be incorporated.

† Minimum for enterotoxin production » pH 5.2, under aerobic conditions.

** Range from 0.86 - >0.99

x Possibly as low as 0.91 or less.

INTOXICATIONS
Food-poisoning characteristics

	Incubation* time	Symptoms*	Duration*	Mortality rate	Infective dose
Cl. botulinum	12-36 h	Impaired vision, dryness of mouth, nausea, vomiting paralysis†		Formerly 30-65%, now much lower (<10%)	0.005-0.5 mg of toxin
Staph. aureus	1-7 h	Nausea, vomiting, abdominal pain, diarrhoea		Rare	10^5+ cells needed to produce toxin <1 mg of toxin
B. cereus					
emetic	1-5 h	Nausea, vomiting	6-24 h	Nil	10^5+ cells (+ toxin)
diarrhoeal	8-16 h	Abdominal pain	12-24 h		10^5+ cells

* Typical - can vary widely

† Affects the nervous system

Methods for Detecting/Counting Foodborne Pathogens in Foods

The sections in Micro-Facts that describe different foodborne pathogens include certain references to methods for their isolation. These references have been selected, in the main, on the basis that they either describe well-verified and established methods, or they review a number of methods. The decision to exclude most references to 'rapid methods' - other than a few selected articles and reviews/books - was a conscious one, and made after some deliberation. As this field is still developing at such a fast rate, and because the choice of a suitable rapid method is one that needs to be taken with regard to costs, facilities available, etc., the vast majority of references have been omitted here. However, the brief list of references on the subject of rapid methods has been extended from the previous edition and is given at the end of this section.

The selection and use of detection methods - whether traditional or 'rapid' - should be carefully considered. Microbiological testing alone - especially end-product testing - cannot ensure food safety (you cannot 'test in' safety), and should be done as part of a HACCP scheme. Unfortunately, many people still seem to feel that end product testing can be a useful means of determining overall safety of food products, coupled with 'positive release'.

Before embarking on any microbiological examination of foods, the questions should be asked: "What are we testing for and why? What will the results mean? What action will need to be taken on finding a particular result?"

Account also needs to be taken of the uneven distribution of micro-organisms in non-liquid foods, and appropriate sampling plans should be used (see, for example, the sampling plans described by the International Commission on Microbiological Specifications for Foods (ICMSF) 1986).

Furthermore, the detection of low numbers of foodborne pathogens may require the careful selection of appropriate resuscitation, enrichment, isolation and confirmatory tests. Pathogens may be damaged, and may be greatly outnumbered by other, competitor, organisms. These facts need to be taken into account when considering methods to be used.

Many standard methods are described by the British Standards Institution (BSI) and the International Standards Organization (ISO) (available through BSI). Other standard methods, relating to dairy products or chocolate and confectionery, are available from the International Dairy Federation (IDF) (through the United Kingdom Dairy Association), or the International Office of Cocoa, Chocolate and Sugar Confectionery (IOCCC), respectively (see chapter on Addresses of Authorities/Sources of Further Information, Etc.). The ICMSF and the Association of Official Analytical Chemists (AOAC) also describe recommended/standard isolation procedures. Certain books, such as 'The Microbiological Safety and Quality of Food: Volume 2' (by Lund, Baird-Parker & Gould), and 'Foodborne Pathogens: An Illustrated Text' (by Varnam & Evans), review or describe many methods for various pathogens, and major media suppliers provide useful guidance on methods, through their catalogues and their technical support.

American standard methods are described in the FDA Bacteriology Analytical Manual (BAM) (see * below) and/or in the APHA Compendium of Methods for the Microbiological Examination of Foods (see + below).

Advice concerning the use of different isolation procedures, whether 'traditional' or rapid, can be obtained through Leatherhead Food International and other research associations.

Bibliography

Authoritative texts on (more traditional) methods

Doyle M.P., Beuchat L.R. Food Microbiology: Fundamentals and Frontiers: 3rd Edition. Washington DC. ASM Press, 2007.

Roberts D., Greenwood M. Practical Food Microbiology. 3rd Edn. Oxford. Blackwell Publishing, 2003.

British Standards Institution. Method for microbiological examination of food and animal feeding stuffs. Part 0. General Laboratory Practices. BS 5763-0:1996. ISO 7218:1996 (Incorporating Amendment No. 1.). British Standards Institution, 2002.

International Commission on Microbiological Specifications for Foods. Microorganisms in Foods 7. Microbiological testing in food safety management. New York. Kluwer Academic/Plenum Publishers, 2002.

American Public Health Association. Compendium of Methods for the Microbiological Examination of Foods. Eds Downes F.P., Ito K. 4th Edn. Washington DC. APHA, 2001.

Andrews W.H. Microbiological methods. Official methods of analysis of AOAC International, volume 1: agricultural chemicals, contaminants, drugs. AOAC International, Ed. Horwitz W. 17th Edn. Gaithersburg. AOAC International, 2000.

Harrigan W.F. Laboratory Methods in Food Microbiology. 3rd Edn. London. Academic Press, 1998.

Collins C.H., Lyne P.M., Grange J.M. Collins and Lyne's Microbiological Methods. 7th Edn. Oxford. Butterworth-Heinemann Ltd, 1995.

*Food and Drug Administration. Bacteriological Analytical Manual (including revisions to 2001). 8th Edn. Gaithersburg. AOAC International, 1995.

Steering Group on the Microbiological Safety of Food, Ministry of Agriculture, Fisheries and Food Methods for Use in Microbiological Surveillance. London. MAFF, 1994.

Food and Agriculture Organization. Food and Nutrition Paper 14-4, Rev. 1. Manual of Food Quality Control, Volume 4: Microbiological Analysis. Eds Food and Agriculture Organization, W. Andrews. Rome. FAO, 1992.

Varnam, A.H., Evans, M.G. Foodborne Pathogens: An Illustrated Text. London. Wolfe Publishing Ltd, 1991.

Health and Welfare Canada, Health Protection Branch. Compendium of Analytical Methods, Volume 3: Laboratory Procedures of Microbiological Analysis of Food. Morin Heights. Polyscience Publishers, 1991.

Health and Welfare Canada. Compendium of Analytical Methods, Volume 1: Official Methods of Microbiological Analysis of Food. Morin Heights. Polyscience Publishers, 1991.

International Commission on Microbiological Specifications for Food. Microorganisms in Foods, Volume 1: Their Significance and Methods of Enumeration. Toronto. University of Toronto Press, 2nd Rev. Edn, 1988.

International Commission on Microbiological Specifications for Foods. Microorganisms in Foods, Volume 2: Sampling for Microbiological Analysis: Principles and Specific Applications. 2nd Edn. Oxford. Blackwell Scientific Publications, 1986, 206-48.

Rapid methods - overviews and books

Butot S., Putallaz T., Sanchez G. Procedure for rapid concentration and detection of enteric viruses from berries and vegetables. *Applied and Environmental Microbiology*, 2007, (January), 73 (1), 186-192.

Fukushima H., Katsube K., Hata Y., Kishi R., Fujiwara S. Rapid separation and concentration of food-borne pathogens in food samples prior to quantification by viable-cell counting and real-time PCR. *Applied and Environmental Microbiology,* 2007, (January), 73 (1), 92-100.

Doyle M.P., Beuchat L.R. Food microbiology: fundamentals and frontiers: 3rd edition. Washington DC. ASM Press, 2007.

Churruca E., Girbau C., Martinez I., Mateo E., Alonso R., Fernandez-Astorga A. Detection of *Campylobacter* jejuni and *Campylobacter* coli in chicken meat samples by real-time nucleic acid sequence-based amplification with molecular beacons. *International Journal of Food Microbiology,* 2007, (June 10), 117 (1), 85-90.

Jassim S.A.A., Griffiths M.W. Evaluation of a rapid microbial detection method via phage lytic amplification assay coupled with live/dead fluorochromic stains. *Letters in Applied Microbiology*, 2007, (June), 44 (6), 673-8.

Rudi K., Kleiberg G.H., Heiberg R., Rosnes J.T. Rapid identification and classification of bacteria by 16S rDNA restriction fragment melting curve analyses (RFMCA). *Food Microbiology,* 2007, (August), 24 (5), 474-81

Tebbutt G.M. Does microbiological testing of foods and the food environment have a role in the control of foodborne disease in England and Wales? *Journal of Applied Microbiology*, 2007, (April), 102 (4), 883-91.

Matthews K.R. Rapid methods for microbial detection in minimally processed foods, in *Microbial Safety of Minimally Processed Foods*. Eds Novak J.S., Sapers G.M., Juneja V.K. Boca Raton. CRC Press, 2003.

Fung D. Rapid methods and automation in microbiology. *Comprehensive Reviews in Food Science and Food Safety*, 2002, 1 (1), 3-22.

Feng P. Development and impact of rapid methods for detection of foodborne pathogens, in *Food Microbiology: Fundamentals and Frontiers*. Eds Doyle M.P., Beuchat L.R., Montville T.J. 2nd Edn. Washington DC. ASM Press, 2001, 775-96.

Fung D.Y.C. Overview of rapid methods of microbiological analysis, in *Food Microbiological Analysis: New Technologies.* Eds M.L. Tortorello, S.M. Gendel. New York. Marcel Dekker, 1997, 1-25.

Griffiths M.W. Rapid microbiological methods with hazard analysis critical control point. *Journal of AOAC International*, 1997 (November-December) 80 (6), 1143-50.

Stanley P.E., Smither R., Simpson W.J. A Practical Guide to Industrial Uses of ATP-luminescence in Rapid Microbiology. Lingfield. Cara Technology Ltd, 1997.

Tortorello M.L., Gendel S.M. Food Microbiological Analysis: New Technologies. New York. Marcel Dekker, 1997.

van der Zee, H., Huis in't Veld, J.H.J. Rapid and alternative screening methods for microbiological analysis. *Journal of AOAC International*, 1997 (July-August) 80 (4), 934-40.

Giese, J. Rapid microbiological testing kits and instruments. *Food Technology*, 1995, 49 (7), 64-70.

Griffiths, M.W. Rapid methods for assessing microbiological quality of foods. *Journal of Rapid Methods and Automation in Microbiology*, 1995, 3 (4), 291-308.

Swaminathan, B., Feng, P. Rapid detection of foodborne pathogenic bacteria, in *Annual Review of Microbiology, Volume 48*. Eds Ornston L.N. , Balows A., Greenberg E.P. Palo Alto. Annual Reviews Inc., 1994, 401-26.

Patel, P.D. Rapid Analysis Techniques in Food Microbiology. Glasgow. Blackie, 1994.

Further recommended books

Blackburn C. de W., McClure P.J. Foodborne Pathogens: Hazards, Risk Analysis and Control. Cambridge. Woodhead Publishing Ltd, 2002.

Doyle M.P., Beuchat L.R., Montville T.J. Food Microbiology: Fundamentals and Frontiers. 2nd Edn. Washington DC. ASM Press, 2001.

Ray B. Fundamental Food Microbiology. 2nd Edn. Boca Raton. CRC Press, 2001.

Forsythe S.J. The Microbiology of Safe Food. Oxford. Blackwell Science, 2000.

Adams M.R., Moss M.O. Food Microbiology. 2nd Edn. Cambridge. RSC, 2000.

Lund B.M., Baird-Parker T.C., Gould G.W. The Microbiological Safety and Quality of Food, Volume 1. Gaithersburg. Aspen Publishers, 2000.

Lund B.M., Baird-Parker T.C., Gould G.W. The Microbiological Safety and Quality of Food, Volume 2. Gaithersburg. Aspen Publishers, 2000.

Jay J.M. Modern Food Microbiology. 6th Edn. Gaithersburg. Aspen Publishers, 2000.

Hui Y.H., Pierson M.D., Gorham J.R. Foodborne Disease Handbook, Volume 1: Bacterial Pathogens. 2nd Edn. New York. Marcel Dekker, 2000.

Hui Y.H., Sattar S.A., Murrell K.D., Nip W.K., Stanfield P.S. Foodborne Disease Handbook, Volume 2: Viruses, Parasites, Pathogens and HACCP. 2nd Edn. New York. Marcel Dekker, 2000.

Mortimore S., Wallace C. HACCP. A Practical Approach. Oxford. Blackwell Science, 2001.

Robinson R.K., Batt C.A., Patel P.D. Encyclopedia of Food Microbiology. London. Academic Press, 1999, 3 volumes.

International Commission on Microbiological Specifications for Foods. Microorganisms in Foods, Volume 6: Microbial Ecology of Food Commodities. London. Blackie, 1998.

Codex Alimentarius Commission, Food and Agriculture Organization. Food Hygiene: Basic Texts. World Health Organization. Rome, FAO, 1997.

Corry J.E.L., Curtis G.D.W., Baird R.M. Progress in Industrial Microbiology, Volume 34: Culture Media for Food Microbiology. Amsterdam. Elsevier, 1995.

International Commission on Microbiological Specifications for Foods. Microorganisms in Foods, Volume 5: Microbiological Specifications of Food Pathogens. London. Blackie, 1996.

Hayes P.R. Food Microbiology and Hygiene. 2nd Edn. London. Elsevier Applied Science Publishers, 1992.

Tu A.T. Handbook of Natural Toxins, Volume 7: Food Poisoning. New York. Marcel Dekker, 1992.

Shapton D.A., Shapton N.F. Principles and Practices for the Safe Processing of Foods. Oxford. Butterworth-Heinemann Ltd, 1991.

Microbial Ecology of International Commission on Microbiological Specifications for Foods. Foods, Volume 1: Factors Affecting the Life and Death of Microorganisms. London. Academic Press, 1980.

Please note: There is no shortage of literature relating to food-poisoning bacteria! For Leatherhead Food International Members, access to such material can be gained through our database ‘Foodline Science’ and LFI’s Regulatory and Technical Consultancy Services Department. Members can also access the extensive collection of books and journals in the Leatherhead Food International library, and utilise the rapid document delivery service now available.

A Few Words about the Bibliographies

The choice of literature to include in these bibliographies was not always an easy one. For the purposes of brevity, some deliberate bias was shown in certain cases to UK and USA publications and to those that are of more practical (rather than academic) interest. The authors apologise for any inevitable important omissions.

BACILLUS CEREUS

Bacillus cereus is now recognised as a significant cause of foodborne illness in humans. It can cause two distinct forms of food poisoning. These are the so-called "diarrhoeal" type and the "emetic" (vomiting) type, caused by the formation of toxins. The emetic type is considered to be most common in the UK.

The Organism *Bacillus cereus*

Bacillus cereus is a large Gram-positive, endospore-forming, motile rod-shaped bacterium with peritrichous flagellae belonging to the family Bacillaceae. It grows best in the presence of oxygen, but also grows well anaerobically. Sporulation readily occurs only in the presence of oxygen. It tends to grow in long chains. The cell width is ≥ 0.9 µm and it produces central to terminal ellipsoid or cylindrical spores that do not distend the sporangia. Under anaerobic conditions both growth and toxin production are reduced.

The genus *Bacillus* encompasses a wide variety of species and strains. Although other *Bacillus* species have been associated with food poisoning, *Bacillus cereus* is the species of most importance to food microbiologists. These other species are covered in more detail at the end of this section.

B. cereus Food Poisoning

Bacillus cereus has been known to be a cause of food poisoning since the early 1950s. *B. cereus* food poisoning occurs after the ingestion of foods in which the organism has grown and formed its toxin(s). Emetic food poisoning arises from the ingestion of emetic toxin that has been pre-formed in the food (i.e. it is intoxication). The emetic toxin has been characterised as a ring form peptide (cereulide) and is thought to be associated with sporulation (1). The less common diarrhoeal type of *B. cereus* poisoning arises from the formation and release of enterotoxin in the small intestine, although the enterotoxin can also be pre-formed in food (i.e. it is an intermediate type of food poisoning) (2). At least two different enterotoxins have been found, but their role in food poisoning is unclear.

Incubation time

The emetic type of illness, characterised by nausea and vomiting has a short onset period of between 1 and 5 h. The diarrhoeal illness, characterised by watery diarrhoea, typically has a longer incubation time of approximately 8-16 h (more usually 10-12 h) (3,4).

Symptoms

Emetic food poisoning symptoms, which usually last for between 6 and 24 h (3,4), include nausea, vomiting and general malaise, and occasionally diarrhoea. This type of food poisoning closely resembles that of *S. aureus*.

Predominant symptoms of diarrhoeal food poisoning are diarrhoea and abdominal cramps, with only occasional nausea and vomiting. Symptoms usually persist for between 12 and 24 h (3,4). However, body aches, fever, chills, and longer incubation and duration times were reported in a large outbreak in the USA (5). This type of food poisoning resembles that of *Clostridium perfringens*.

Mortality

Although recovery from *B. cereus* food poisoning is normally within 24 hr with no complications, there have been reported fatalities due to ingestion of emetic toxin (6).

Infective dose

Large numbers of *B. cereus* are required to cause food poisoning. Numbers found in foods implicated in *B. cereus* poisoning are typically within the range of 10^5-10^9/g, although smaller numbers have occasionally been reported. *B. cereus* food poisoning may also occur after ingestion of food containing the pre-formed heat-stable emetic toxin, but which, as a result of further processing or re-heating, contains few viable organisms (1,5). Any food containing more than 10^3 *B. cereus* /g cannot be considered completely safe for consumption.

Foods involved

Emetic food poisoning is almost exclusively associated with farinaceous foods, especially boiled or fried rice, and cooked pasta and noodle dishes. Most of the outbreaks in the UK have been associated with cooked rice from Chinese restaurants and take-away outlets. In Chinese restaurants, cooked rice (prepared in bulk) is allowed to cool at room temperature overnight, after which it may be reheated/fried for serving. These practices result in the survival and growth of strains of *B. cereus*, producing the pre-formed toxin in the food (2,7).

The diarrhoeal type of *B. cereus* food poisoning has been associated with a wide range of proteinaceous foods, most commonly meat or vegetable dishes, soups, sauces and puddings. In general, this type of food poisoning occurs when spores survive the cooking process, and unsatisfactory post-cooking storage conditions permit the spores to germinate and growth to occur (2).

Incidence of *B. cereus* Food Poisoning

The worldwide reported incidence of *B. cereus* food poisoning ranges from <1-22% of total food-poisoning outbreaks (<1-18% of cases) of known bacterial origin, according to country (2).

In the US during 1988-1992, 21% of food poisoning outbreaks were due to *B. cereus*. On average, about 14% of food poisoning outbreaks in Canada are associated with *B. cereus* per year. In the UK, during 1989-1993, 68 % of food poisoning outbreaks were due to *B. cereus* food poisoning (9). Because of the relatively mild and short-lived nature of the illness, a large proportion of food-poisoning incidents caused by this organism may go unreported.

It has been estimated that there are 84,000 cases of *B. cereus* food poisoning in the US each year, with an annual cost of USD36m. Canada has an estimated 23,000 cases at a cost of USD10m (9).

Sources

Humans

Bacillus species are often present in low numbers in normal stools as transient flora, which reflects dietary intake (low numbers of *B. cereus* can be ingested with no ill effect). In cases of food poisoning, the organism can be excreted in fairly high numbers for up to 48 h after onset (7,10).

Animals and environment

Bacillus cereus is widely distributed in nature. It is common in a wide range of environments, including soils (especially rice paddy soil), sediments, dust, natural water, animal hair and vegetation.

Foods

Because the *Bacillus cereus* group is ubiquitously distributed in the environment, it can easily contaminate various types of foods, especially products of plant origin. It is common in many foods, notably cereals and cereal derivatives, as well as spices, meat and meat products, pasteurised liquid egg, rice, ready-to-eat vegetables, milk and dairy products, and dried foods. Its association with milk, where it can be a significant cause of spoilage in raw and heat-treated products (especially 'bitty cream'), as well as with rice, may be a particular cause for concern. However, proper cold storage appears to be effective in preventing spore outgrowth and toxin production.

Growth/Survival Characteristics of the Organism in Foods

Temperature

Typically, the vegetative cells of *B. cereus* have an optimum growth temperature of 30-35 °C, and a maximum ranging from 48-55 °C (4,11,12). However, psychrotrophic strains have been identified - especially in milk and dairy products - capable of growing within the temperature range 4-37 °C (13). In addition, most of the psychrotrophic strains tested have been shown to produce enterotoxin (14) and production at temperatures as low as

4 °C has been reported, although concerns have been raised about the methodology used in these studies (12). The emetic toxin has been shown to be produced optimally at 25-30 °C (12).

Heat resistance

Vegetative cells of *B. cereus* are readily destroyed by pasteurisation or equivalent heat treatments. However, spores can survive quite severe heat processes. *B. cereus* spores from different strains show varying heat resistance; D_{95}-values of between 1.2 and 36 min have been reported (2). It has been shown that strains commonly implicated in food poisoning are more heat-resistant than other strains, and are therefore more likely to survive cooking. The heat resistance of spores is also enhanced in low a_w or high fat foods.

The diarrhoeal enterotoxin is heat-sensitive, being destroyed at 56 °C for 5 min (2,4). However, the emetic toxin is extremely heat-resistant, and can withstand heating at 126 °C for 90 min (2,12). This heat resistance is very significant in the context of food safety, where food, especially rice, is prepared in advance and left at room temperatures before reheating. This practice can allow the growth of *B. cereus* and formation of heat-stable toxin, which is not affected by reheating of the food (such as may occur in the preparation of fried rice in restaurants).

pH

B. cereus has been reported to be capable of growth at pH values between 4.3 and 9.3, under otherwise ideal conditions (2,10).

The emetic toxin is stable in the pH range 2-11 (4), whereas the diarrhoeal enterotoxin is unstable at pH values of <4 or >11 (2).

Water activity/salt

In the presence of NaCl as a humectant, *B. cereus* will not grow at a_w of 0.93. However, growth was found at a_w of 0.93 (but not at a_w of 0.92) when glycerol was used as a humectant (8). As fried rice is known to support the growth of *B. cereus*, and the water activity of fried rice may be as low as 0.91, this may be a more reliable guide to its minimum (2). *B. cereus* can tolerate salt concentrations up to 7.5% (4).

Atmosphere

Most studies have shown *B. cereus* to be sensitive to the antimicrobial effects of carbon dioxide (CO_2). It affects both spore germination and growth of the organism. Under 100% CO_2, *B. cereus* growth is only 17% of that observed under aerobic conditions, and 33% of that under anaerobic conditions (15).

Summary of Control of *B. cereus* in Foods

Because *B. cereus* is ubiquitous and has the ability to form spores, it ensures some degree of contamination of the food chain and survival through all stages of food processing short of retorting. As already mentioned, rice, in particular, can harbour *B. cereus*.

The ingestion of low levels of the organism does not represent a hazard. However, efforts must be made to minimise the germination and outgrowth of spores that survive thermal treatment. Primary control of this type of intoxication consists in prevention of time and temperature abuse, particularly in cooked products. Cooked foods should be held hot before consumption (minimum 63 °C), or, if such foods are to be stored, they must be cooled rapidly and kept at refrigeration temperatures for a limited time, unless the pH and/or a_w of the food are such that growth is prevented.

OTHER *BACILLUS* SPECIES

Species of *Bacillus* other than *B. cereus* have been shown to be responsible for incidents of food poisoning. These species include *B. licheniformis* and *B. subtilis*. In comparison with *B. cereus*, the accumulation of knowledge about these organisms in relation to food safety is in its infancy. Foodborne incidents involving these bacteria, however, have occasionally been reported.

B. subtilis has been implicated in food poisoning caused by 'ropy' bakery products, as well as a number of other foods, such as custard powder, synthetic fruit drinks, mayonnaise, meat, seafood with rice and pastry. *Bacillus subtilis* food poisoning has a rapid onset (median 2.5 h) and the main symptom is acute vomiting, with diarrhoea commonly following. *B. licheniformis* food poisoning is also associated with cooked meat and vegetable dishes, as well as meat or vegetable stews, or curried meat or poultry with rice. *Bacillus licheniformis* has a slightly longer onset (median 8 h), and the main feature is diarrhoea, although vomiting occurs in about half the cases. This type of food poisoning closely resembles that caused by *Cl. perfringens*.

B. subtilis has been implicated in food poisoning more often than *B. licheniformis*. The Health Protection Agency stated that *B. subtilis* was implicated in a total of 70 outbreaks reported to the Food Hygiene Laboratory during the period 1974-89. In the same period, *B. licheniformis* was implicated in a total of 32 outbreaks (2,7). In general, where either of these bacteria has been implicated in food poisoning incidents, the contamination level in food is high (10^5/g or more) (2,7).

It is currently very difficult to differentiate *B. cereus* and *B. thuringiensis* by conventional methodology. However, since *B. thuringiensis* is now widely used in agriculture for its insecticidal properties, it is likely to occur more frequently in some food materials in future. Therefore there is a need for better identification methods to be developed. Furthermore, there have been reports of enterotoxin production by *B. thuringiensis* strains isolated from food (16).

It can be concluded that the presence of large numbers of *Bacillus* in any food should be viewed with suspicion.

Bibliography

References

1. Granum P.E. *Bacillus cereus* in *Foodborne Pathogens Microbiology and Molecular Biology*. Eds Fratamico P.M., Bhunia A.K., Smith J.L. Great Britain. Caister Academic Press, 2005, 409-20.

2. Kramer J.M., Gilbert R.J. *Bacillus cereus* and other *Bacillus* species. *Foodborne Bacterial Pathogens*. Ed. Doyle M.P. New York. Marcel Dekker, 1989, 21-70.

3. Bennett R.W. *Bacillus cereus* in *Guide to Foodborne Pathogens*. Eds Labbé R.G., Garcia S. New York. A John Wiley & Sons, Inc., Publications, 2001, 51-9.

4. Rajkowski K.T., Bennett R.W. *Bacillus cereus* in *International Handbook of Foodborne Pathogens*. Eds Miliotis M.D., Bier J.W. New York. Marcel Dekker, Inc., 2003, 27-40.

5. Luby S., Jones J., Dowda H., Kramer J., Horan J. A large outbreak of gastroenteritis caused by diarrhoeal toxin producing *Bacillus cereus*. *Journal of Infectious Diseases*, 1993, 167, 1452-5.

6. Dierick K., Van Coillie E., Swiecicka I., Meyfroidt G., Devlieger H., Meulemans A., Hoedemaekers G., Fourie L., Heyndrickx M., Mahillon J. Fatal Family Outbreak of *Bacillus cereus* Associated Food Poisoning. *Clinical Microbiology*, 2005, 43(8), 4277–9.

7. Lund, B.M. Foodborne disease due to *Bacillus* and *Clostridium* species. Lancet, 1990, 336 (8721), 982-6.

8. Schraft H., Griffiths M.W. *Bacillus cereus* gastroenteritis in *Foodborne infections and intoxications*. Eds Riemann H.P., Cliver D.O. London. Academic Press, 2005, 3rd edition, 563-82.

9. Schultz, F.J., Smith, J.L. *Bacillus*: recent advances in *Bacillus cereus* food poisoning research. *Foodborne Disease Handbook, Volume 1: Diseases Caused by Bacteria*. Eds Hui Y.H., Gorham J.R., Murrell K.D., Cliver D.O. New York. Marcel Dekker, 1994, 29-62.

10. Fermanian C., Fremy J.-M., Lahellec C. *Bacillus cereus* pathogenicity: a review. *Journal of Rapid Methods and Automation in Microbiology*, 1993, 2 (2), 83-134.

11. Fermanian C., Fremy J.M., Claisse M. Effect of temperature on the vegetative growth of type and field strains of *Bacillus cereus*. *Letters in applied Microbiology*, 1994, 19 (6), 414-8.

12. International Commission on Microbiological Specifications for Foods. *Bacillus cereus*, in *Microorganisms in Foods, Volume 5: Microbiological Specifications of Food Pathogens*. Ed. International Commission on Microbiological Specifications for Foods. London. Blackie, 1996, 20-35.

13. van Netten P., van de Moosdijk A., van Hoensel P., Mossel D.A.A., Perales I. Psychrotrophic strains of *Bacillus cereus* producing enterotoxin. *Journal of Applied Bacteriology*, 1990, 69 (1), 73-9.

14. Dufrenne J., Soentoro P., Tatini S., Day T., Notermans S. Characteristics of *Bacillus cereus* related to safe food production. *International Journal of Food Microbiology*, 1994, 14 (2), 87.

15. Farber J.M. Microbiological aspects of modified atmosphere packaging technology - a review. *Journal of Food Protection*, 1991, 54 (1), 58-70.

16. Damgaard P.H., Larsen H.D., Hansen B.M., Bresciani J., Jorgensen K. Enterotoxin-producing strains of *Bacillus thuringiensis* isolated from food. *Letters in Applied Microbiology*, 1996, 23 (3), 146-50.

Further reading

Cramer M.M. Microorganisms of concern for food manufacturing, in *Food plant sanitation: design, maintenance and good manufacturing practices*. Ed. Cramer M.M. Boca Ranton. CRC Press, 2006, 53-74.

Lee S.Y., Chung H.J., Shin J.H., Dougherty R.H., Kang D.H. Survival and growth of foodborne pathogens during cooking and storage or Oriental-style rice cakes. *Journal of Food Protection*, 2006, 69 (12), 3037-42.

den Besten H.M.W., Mataragas M., Moezelaar R., Abee T., Zwietering M.H. Quantification of the effects of salt stress and physiological state on the thermotolerance of *Bacillus cereus* ATCC 10987 and ATCC. 14579. *Applied and Environmental Microbiology*, 2006, 72 (9), 5884-94.

Scorch K.J., Robertson R.E., Craven H.M., Pearce L.E., Szabo E.A. Inactivation of *Bacillus* spores in reconstituted skim milk by combined high pressure and heat treatment. *Journal of Applied Microbiology*, 2006, 101 (1), 172-180.

Moussa-Boudjemaa B., Gonzalez J., Lopez M. Heat resistance of *Bacillus cereus* spores in carrot extract acidified with different acidulants. *Food Control*, 2006, 17 (10), 819-24.

Armstrong G.N., Watson I.A., Stewart-Tull D.E. Inactivation of *Bacillus cereus* spores on agar, stainless steel or in water with a combination of neodymium:YAG laser and UV irradiation. *Innovative Food Science and Emerging Technologies*, 2006, 7(1-2), 94-9.

Carlin F., Fricker M., Pielaat A., Heisterkamp S., Shaheen R., Salkinoja Salonen M., Svensson B., Nguyen-the C., Ehling-Schulz M. Emetic toxin-producing strains of *Bacillus cereus* show distinct characteristics within the *Bacillus cereus* group. *International Journal of Food Microbiology*, 2006, 109 (1-2), 132-8.

Grande M.J., Lucas R., Abriouel H., Valdivia E., Omar N.B., Maqueda M., Martinez-Bueno M., Martinez-Canamero M., Galvez A. Inhibition of toxicogenic *Bacillus cereus* in rice-based foods by enterocin AS-48. *International Journal of Food Microbiology*, 2006, 106 (2), 185-94.

Nguyen-the C., Broussolle V. *Bacillus cereus*: factors affecting virulence, in *Understanding pathogen behaviour: virulence, stress response and resistance*. Ed. Griffiths M. Cambridge. Woodhead Publishing Ltd, 2005, 309-30.

Granum, P.E. *Bacillus cereus*, in *Food Microbiology: Fundamentals and Frontiers. 2nd Edn*. Eds Doyle M.P., Beuchat L.R., Montville T.J. Washington DC.ASM Press, 2001, 373-81.

Granum P.E., Baird-Parker T.C. *Bacillus* species. *The microbiological safety and quality of food, volume 2*. Eds Lund B.M., Baird-Parker T.C., Gould G.W. Gaithersburg. Aspen Publishers, 2000, 1029-39.

International Commission on Microbiological Specifications for Foods. *Bacillus cereus*. Microorganisms in Foods, Volume 5: Microbiological Specifications of Food Pathogens. Ed. International Commission on Microbiological Specifications for Foods. London. Blackie, 1996, 20-35.

Bacillus cereus: Management of outbreaks of foodborne illness. *Department of Health Working Group*. London. HMSO, 1994, 58-9.

International Dairy Federation. *Bacillus cereus* in milk and milk products. IDF Bulletin No. 275, Brussels: IDF, 1992.

Methods of detection

British Standards Institution. Microbiology of food and animal feeding stuffs. Horizontal method for the determination of low numbers of presumptive *Bacillus cereus*. Most probable number technique and detection method. BS EN ISO 21871:2006.

Bennett R.W., Belay N. *Bacillus cereus*. *Compendium of Methods for the Microbiological Examination of Food*. Eds American Public Health Association, Downes F.P., Stern N.J. Washington DC. APHA, 2001, 311-16.

Christiansson A., Te Giffel M..C., Notermans S.H.W., Beumer R.R., Griffiths M.W. Taxonomy and identification of *Bacillus* species. *Bacillus cereus*: its toxins and their significance. *Bacillus cereus* in milk and milk products: advances in detection, typing and epidemiology. *Bulletin of the International Dairy Federation, No. 357*. Ed. International Dairy Federation. Brussels. IDF, 2000, 40-54.

Bennett, R.W. *Bacillus cereus* diarrheagenic enterotoxin. *Bacteriological Analytical Manual. 8th Edn*. Ed. Food and Drug Administration. Gaithersburg. AOAC International, 1995.

Rhodehamel E.J., Harmon S.M. *Bacillus cereus*. *Bacteriological Analytical Manual. 8th Edn*. Ed. Food and Drug Administration. Gaithersburg. AOAC International, 1995.

Van Netten P., Kramer J.M. Media for the detection and enumeration of *Bacillus cereus* in foods. *Culture Media for Food Microbiology*. Eds Corry J.E.L., Curtis G.D.W., Baird R.M. Amsterdam. Elsevier, 1995, 35-49.

Buchanan R.L., Schultz F.J. Comparison of the Tecra VIA kit, Oxoid BCET-RPLA kit and CHO cell culture assay for the detection of *Bacillus cereus* diarrhoeal enterotoxin. *Letters in Applied Microbiology*, 1994, 19 (5), 353-6.

Food and Agriculture Organization, Andrews, W. *Bacillus cereus,* in *Manual of Food Quality Control, Volume 4: Microbiological Analysis*. Eds Food and Agriculture Organization, Andrews W. Rome. FAO, 1992, 199-206.

CAMPYLOBACTER

Campylobacter fetus was first isolated from aborted sheep fetuses, and it was initially called *Vibrio fetusovid* because of the curved shape. It was later known by various names, for example, *Vibrio fetus* - isolated from aborted cattle, *Vibrio jejuni* - cause of winter dysentery outbreaks in cattle. Finally, the species was given the new genus name *Campylobacter*, meaning curved rod in Greek (1). Its role as a human pathogen was not appreciated until 1977, when suitable methods were developed for its isolation. Since then, it has become increasingly recognised as an important cause of gastroenteritis.

The Organism *Campylobacter*

Campylobacter are Gram-negative, oxidase and catalase-positive, non-spore -forming rods with a characteristic spiral (s-shape) morphology belonging to the family Campylobacteriaceae. Most *Campylobacter* species are motile (corkscrew-like darting motility) and produce a single polar flagellum. They are microaerophilic, requiring special atmospheric conditions for growth. For these reasons they are not normally capable of growing in foods. This distinguishes them from other food-poisoning bacteria, but despite this fact, *Campylobacter* is the leading cause of bacterial gastrointestinal illness in the UK (2). Because of their small size, they can be separated from most other Gram-negative bacteria by use of a 0.65 μm filter (3).

The genus *Campylobacter* currently consists of 16 species and 6 sub-species (1,4). Some former species of *Campylobacter* have been placed in the genus *Arcobacter* and some are now classified as species of *Helicobacter*.

Campylobacter Enteritis

Campylobacter enteritis can be caused by a number of species, namely *C. jejuni*, *C. coli*, *C. lari*, *C. fetus* and *C. upsaliensis* (1). *C. jejuni* and *C. coli* are the most common species associated with diarrhoeal illness and produce indistinguishable clinical symptoms. *C. jejuni* is the most prevalent in the UK, accounting for 92% of all cases (2).

The illness may start with cramping abdominal pain and diarrhoea. Typical symptoms include fever, chills, headache, myalgia and occasionally delirium (1). It is a more intense and long-lasting abdominal pain than other foodborne illnesses, and the occasional presence of blood or mucus in the stools may be cause for alarm for sufferers. This may explain the large number of cases that are reported to doctors, in contrast to the milder, shorter-lived symptoms caused by other food-poisoning organisms such as *Cl. perfringens* or *B. cereus*.

For patients who require antimicrobial therapy, increasing resistance to fluoroquinolones has become a problem in most parts of the world (4).

Incubation time

The onset of illness caused by *Campylobacter* occurs between 1 and 11 days after exposure, usually 2.5 days (5).

Symptoms

The illness caused by *Campylobacter* is not easily distinguished from other types of gastrointestinal disease. Symptoms vary widely; they can be mild, with little obvious sign of illness, or quite severe, and usually last for between 1 and 7 days, but sometimes last for several weeks. Almost all cases experience diarrhoea, which may be profuse, lasting for 1-3 days, and sometimes involves the excretion of blood and mucus in the stools. Abdominal pain can persist for up to 7 days and recurrence of symptoms can occur. Nausea is common, but vomiting is uncommon (1,6,7).

The disease is usually self-limiting, but complications such as appendicitis, meningitis and reactive arthritis occur in about 1% of patients, 1-2 weeks after onset of the illness. Bacteraemia and Guillain-Barré syndrome are also possible complications (2,7).

Mortality

Almost all cases of *Campylobacter* enteritis recover from illness with a week; although there may be complications, death from *Campylobacter* enteritis is very rare. Mortality associated with infection maybe underestimated, however, and in the US, it has been estimated that the case fatality rate is about 124 deaths per annum (1,7).

Infective dose

The infective dose for *Campylobacter* enteritis is thought to be low; infection has been established with as few as 50-500 organisms. The fact that *Campylobacter* cannot normally grow in foods does not necessarily limit its ability to cause infection (1).

Foods involved

Because the vast majority of reported cases of *Campylobacter* enteritis are sporadic rather than associated with outbreaks, food vehicles are difficult to identify. Furthermore, *Campylobacter* is not often fully identified, and only rarely serotyped (about 15% of isolates from reports to CDSC in 1989/91 were speciated). Consequently, in contrast to salmonellosis, where specific serotypes of *Salmonella* can be traced to specific foods, *Campylobacter* infections are more difficult to trace to a specific food vehicle (8,9,10).

Where a food vehicle has been implicated, *Campylobacter* enteritis has been most commonly associated with undercooked poultry, with cross-contamination from raw poultry to other foods that are not subsequently cooked, and with raw or bird-pecked milk (2). Large outbreaks have been associated with raw and inadequately pasteurised milk and contaminated water supplies (5).

Incidence of *Campylobacter* Food Poisoning

Campylobacter enteritis is prevalent worldwide, and is recognised as an important source of diarrhoeal illness in the public health communities. In Europe and North America it is the number one bacterial cause of gastro intestinal disease (11,12). The increase in the number of reports of *Campylobacter* infections may be due to a combination of factors, including increased surveillance, changes or improvements in culture and molecular methods, and the recognition that they cause primarily sporadic rather than outbreak disease (11).

A study of the levels of *Campylobacter* incidence in England and Wales (1990-1999) revealed an average annual rate of 78.4 +/- 15 cases (11). In 1997, there were approximately 50,000 laboratory reported infections in England and Wales (source: PHLS). This represents an annual reporting rate of about 900/million of the population. However, in the same year only 10 general outbreaks were recorded, emphasising the sporadic nature of disease caused by *Campylobacter*.

Based on active surveillance data collected through CDC's FoodNet, it is estimated that 1% of the USA population develops *Campylobacter* infection each year. In the USA, gastroenteritis caused by *Campylobacter* peaked in 1997, declining to an infection rate of 13.8/100,000 of the population by 2001. Likewise, the incidence of *Campylobacter* infections in the USA appears to be decreasing – only 12.6 incidences per 100,000 persons were reported in 2003 as compared to 21.7 during 1996-1998 (4).

The greatest incidence of *Campylobacter* infection is seen in infants and young adults (peak age group is 10-29 years), especially young males (1). There is also a seasonal trend in incidence - the highest incidence occurs in spring/early summer (generally between mid-June and mid-July). This peak could be due to informal eating outside, such as barbecues, coupled with an increase in temperature. There is some evidence linking it to agricultural activities (11). There is also more infection amongst the rural community than in urban populations.

As in the case of other forms of foodborne infection, these reported isolations probably underestimate the true incidence of *Campylobacter* infection. One small study in the early 1980s indicated that the true incidence might be over half a million cases each year in England and Wales. Another study suggested that cases might exceed one million annually (13).

Sources

Humans

Faecal-oral person-to-person transmission of infection has been reported for *C. jejuni*, but it is uncommon. This type of transmission can occur if personal hygiene is poor. Asymptomatic carriers are also reported to occur, but are rare in developed countries (14). However, humans may act as vectors and transfer the organism into poultry production areas on contaminated clothing and footwear.

Animals and environment

Campylobacter is not an environmental organism but rather a zoonotic organism; it can be found in the intestinal tract of a wide range of warm-blooded animals. It is especially common in birds, which relates to the organism's high optimum growth temperature. Wild birds are considered to be an important reservoir of infection for domestic and food animals. Poultry is also a natural host for *Campylobacter* (2).

In addition to birds, campylobacters may be found in the intestinal tract of cattle, sheep, pigs, goats and domestic pets, especially puppies and kittens. *Campylobacter jejuni* has also been isolated from houseflies (15).

Water can become contaminated with *Campylobacter* and can act as a vehicle for *Campylobacter* infections - either directly or indirectly. There have been several small waterborne outbreaks in the UK, and large waterborne outbreaks have been reported in other countries. These have mainly been associated with defective chlorination systems. *Campylobacter* may enter a viable but non-culturable state in water (although this is disputed by some researchers) (7,16).

Isolation of *Campylobacter* species has also been made from mud and sewage.

Foods

Poultry - either undercooked, or as a source of contamination of other foods, such as salads in the kitchen - is commonly considered to be the major cause of *Campylobacter* infections in man (2,6,16).

Cross-contamination during processing of poultry can lead to a contamination rate of *Campylobacter* in processed poultry of 60% or more, with numbers per carcass ranging from 10^2-10^4 cfu/g (3). The contamination rate of fresh poultry is higher than in frozen poultry, and it has been suggested that the increase in the reported incidence of *Campylobacter* infection over recent years may be linked to increased consumption in the UK of fresh (as opposed to frozen) poultry. In a Food Standards Agency study carried out in 2001, the prevalence of *Campylobacter* in UK poultry was found to be 50%, but the contamination rate in fresh poultry samples was 63%, with that in frozen poultry being only 33%. Studies in Spain, Japan and Germany have produced similar results, reporting contamination rates of 49.5%, 45.8%, and 41.1% respectively (17,18,19).

Although *Campylobacter* can be isolated readily from fresh red meat carcases, air and drying of the surface lead to poor survival, such that red meat is less prone to being contaminated than poultry. Reported incidence of *Campylobacter* spp. ranges from 0-23.6% in beef, from 0-15.5% in lamb, and from 1-23.5% in pork (20). *C. coli* tend to be isolated from pork more frequently than *C. jejuni*.

Milk borne *Campylobacter* infections are still a problem in England and Wales, where the sale of raw milk is still permitted, although declining in importance. This is in contrast to Scotland, where the incidence of outbreaks from *Campylobacter* infections associated with milk has seen a major decline since the pasteurisation of milk was made compulsory in 1983. Bird-pecked milk bottles are also becoming recognised as a major source of *Campylobacter* infections, especially in the late spring. Recent studies have shown that in some parts of the UK, a high percentage of cases in the early summer peak of *Campylobacter* infections is associated with consumption of milk from bottles pecked by magpies or jackdaws (1,2,16). A recent *Campylobacter* outbreak is thought to be associated with bird-pecked milk (21).

In addition to animal sources, vegetables (e.g. lettuce) and shellfish may also be contaminated with *Campylobacter*, and prevalence in oysters and mussels has been reported as 47-69% and 6-27% respectively (18). More recently, on the basis of epidemiological data, bottled mineral water has been proposed as a significant vehicle for *Campylobacter* infection. More unusual vehicles for foodborne campylobacteriosis includes garlic butter, sweet potato, stir-fried food and tuna salad (2).

Growth/Survival Characteristics of the Organism in Foods

As already indicated, *Campylobacter* is a fragile and fastidious organism. It cannot normally grow in foods (requiring a mammalian or avian host for growth), although it may survive in food. Campylobacters have specific temperature atmosphere requirements, as well as being particularly sensitive to oxygen breakdown products and to drying. Viable, but non-culturable campylobacters have been reported (22).

Temperatures

C. jejuni and *C. coli* can only grow at temperatures above about 30 °C; they (and *C. lari*) are consequently referred to as the thermophilic group of Campylobacters. Their optimum temperature for growth is between 42 °C and 43 °C, with a maximum of 45 °C (6). *Campylobacter* survives poorly at room temperatures (around 20-23 °C) consequently they will not multiply during food processing or food storage.

Chilling is known to promote survival of *C. jejuni*. Conversely, it is generally more sensitive to freezing, although there may be some survival for long periods (1,2,23). Despite its inability to grow at low temperatures *C jejuni* is still metabolically active at temperatures far below its minimal growth temperature and also remains motile. As a consequence of this, the organism is still able to move to favourable environments at temperatures as low as 4 °C (2).

Heat resistance

Campylobacter jejuni is very heat-sensitive. Heat injury can occur at 46 °C or higher; D-values at this temperature (in skim milk) of 7.2-12.8 min have been reported (24). At 55 °C, the range was 0.74-1.0 min (6). The organism cannot survive normal milk pasteurisation. In meat and poultry, heat treatments sufficient to kill *Salmonella* will also readily kill *Campylobacter*. D-values ranging from a few seconds to less than 1 min at temperatures between 57 °C and 60 °C have been reported for *Campylobacter* in meat. Low z-values (*ca* 3.5 °C) have also been demonstrated for *Campylobacter* (25).

pH

The optimum pH for growth is reported to be in the range 6.5 -7.5, and no growth is observed below pH 4.9 (6). The response of *C. jejuni* to pH is influenced by temperature and the type of acid used to adjust the pH. At similar pH values, the organism was most rapidly inactivated at 42 °C, at an intermediate rate at 25 °C, and slowly at 4 °C (25). Lactic acid is more inhibitory than hydrochloric acid at the same pH level (23).

Water activity/salt

Campylobacter is quite sensitive to sodium chloride; levels of 2% or more NaCl can be bactericidal to the organism. The effect is temperature-dependent, and decreases with decreasing temperature (24).

Campylobacter is particularly sensitive to drying; it does not survive well in dry environments. Minimum water activity for growth is 0.98. However, a study in 1998 suggested that *Campylobacter* may be able to survive for some time on wooden cutting boards. It may even be able to grow under certain circumstances if the ambient temperature is high. There is always a potential cross-contamination risk (26).

Atmosphere

Campylobacter is both microaerophilic and 'capnophilic' (liking carbon dioxide). It will grow best in an atmosphere containing 10% carbon dioxide and 5-6% oxygen. Growth is also enhanced by hydrogen. The organism will normally die quickly in the presence of air; it is particularly sensitive to oxygen breakdown products. Vacuum- or gas-packing appears to have little major impact on the survival of *Campylobacter* on chilled meat or poultry (23,24).

Irradiation

Campylobacters are more sensitive to ionising radiation than salmonellae or *Listeria*, and therefore treatments applied to poultry that are designed to eliminate these pathogens are effective against *Campylobacter* (27). D-values of 0.08-0.32 kGy have been reported, but the rate of inactivation is dependent on food type and temperature, with frozen foods requiring more severe treatments than chilled foods (28). Rate of killing is also dependant on the growth phase of the organisms.

Summary of Control of *Campylobacter* in Foods

Since poultry is considered to be one of the most important sources of human infection, prevention of foodborne *Campylobacter enteritis* rests heavily on the control of the organism in poultry. Measures should be taken to minimise the contamination of chickens during production and to minimise the spread of contamination amongst carcases during processing. Several European countries, notably Denmark and Norway, have significantly reduced the prevalence of *Campylobacter* in poultry flocks by improving hygiene in production units and introducing new measures to prevent infection in disease-free flocks. Further control measures involve practicing good food hygiene. For example, prevention of cross-contamination of prepared foods with *Campylobacters* from raw meat (particularly poultry) and adequate refrigeration. Poultry must be thoroughly cooked. The consumption of raw milk and 'bird-pecked' milk also involves a risk of *Campylobacter* infection.

There is a lack of information on the effects of preservatives on *C. jejuni*, but it is likely that preservatives will inhibit the growth of organisms rather than kill it (2).

Bibliography

References

1. Hu L., Kopecko D.J. *Campylobacter* Species, in *International Handbook of Foodborne Pathogens.* Miliotis M.D., Bier J.W. New York. Marcel Dekker Inc., 2003, 181-98.

2. Park S. *Campylobacter*: stress response and resistance, in *Understanding pathogen behaviour: virulence, stress response and resistance.* Ed. Griffiths M. Cambridge. Woodhead Publishing Ltd., 2005, 279-308.

3. Jay J., Loessner M.J., Golden D.A. Foodborne Gastroenteritis Caused by *Vibrio*, *Yersinia*, and *Campylobacter* Species in Modern Food Microbiology. Eds Jay J., Loessner M.J., Golden D.A. USA. Springer, 2005, 668-671.

4. Nachamkin I., Guerry P. *Campylobacter* Infection, in *Foodborne Pathogens Microbiology and Molecular Biology.* Fratamico P.M., Bhunia A.K., Smith J.L. Great Britain. Caister Academic Press, 2005, 285-94.

5. A working part of the PHLS *Salmonella* Committee. The prevention of human transmission of gastrointestinal infections, infestations and bacterial intoxications. *CDR Review,* 1995, 5 (11), R157-72.

6. International Commission on Microbiological Specifications for Foods. Microorganisms in Foods, Vol. 5; in Microbiological Specifications of Food Pathogens. International Commission on Microbiological Specifications for Foods. London: Blackie, 1996, 45-65.

7. Nachamkin I. *Campylobacter jejuni.* Food Microbiology; Fundamentals and Frontiers. Eds Doyle M.P, Beuchat L.R., Montville T.J. Washington DC: ASM Press, 1997, 159-70.

8. Phillips, C.A, Incidence, epidemiology, and prevention of foodborne *Campylobacter* species. *Trends in Food science and Technology,* 1995, 6(3) 83-87.

9. Franco, D.A., Williams, C.E. *Campylobacter jejuni*, in *Foodborne Disease Handbook, Vol. 1: Diseases caused by Bacteria.* Eds Hui Y.H., Gorham J.R., Murrell K.D., Cliver D.O. New York. Marcel Dekker, 1994, 71-96.

10. Sockett, P.N., Cowden, J.M., Le Baigue, S., Ross, D., Adak, G.K., Evans, H. Foodborne disease surveillance in England and Wales: 1989-1991 *CDR Review*, 1993, 3 (12), R159-73.

11. Mandrell R.E., Miller W.G. *Campylobacter,* in *Emerging foodborne pathogens*. Eds Motarjemi Y., Adams M. Cambridge. Woodhead Publishing Ltd., 2006, 476-521.

12. Pebody R.G., Ryan M.J., Wall, P.G. Outbreaks of *Campylobacter* infection: rare events for a common pathogen. *CDR Review*, 1997, 7 (3), R33-7.

13. Pearson A.D., Healing T.D. The surveillance and control of *Campylobacter* infection. *CDR Weekly*, 1992, 2 (12), R133-9.

14. Stern, N.J., Line, J.E. *Campylobacter.* The microbiological safety and quality of food, volume 2. Eds Lund B.M., Baird-Parker T.C., Gould G.W. Gaithersburg. Aspen Publishers, 2000, 1040-56.

15. Rosef O., Kapperud G. House flies (Musca domestica) as possible vectors of *Campylobacter* fetus subsp. jejuni. *Applied and Environmental Microbiology*, 1983, 45 (2), 381-3.

16. Phillips C.A. Incidence of *Campylobacter* and possible modes of transmission. *Nutrition and Food Science,* 1995 (January/February), 12-7.

17. Dominguez C., Gomez I., Zumalacarregui J. Prevalence of *Salmonella* and *Campylobacter* in retail chicken meat in Spain. *International Journal of Food Microbiology*, 2002 (January 30), 72 (1-2), 165-8.

18. Ono K., Yamamoto K. Contamination of meat with *Campylobacter jejuni* in Saitama, Japan. *International Journal of Food Microbiology*, 1999 (March 15), 47 (3), 211-19.

19. Atanassova V., Ring C. Prevalence of *Campylobacter* spp. in poultry and poultry meat in Germany. *International Journal of Food Microbiology*, 1999 (October 15), 51 (2-3), 187-90.

20. Jacobs-Reitsma W. *Campylobacter* in the food supply. *Campylobacter* (2nd Edition). Eds Nachamkin I., Blaser M.J. Washington DC. ASM Press, 2000, 467-81.

21. Stuart J., Sufi F., McNulty C., Park P. Outbreak of *Campylobacter* enteritis in a boarding school associated with bird pecked bottle tops, *CDR Review,* 1997, 7 (3), R38-40.

22. Westfall H.N., Rollins D.M., Weiss E. Substrate utilization by *Campylobacter jejuni* and *Campylobacter coli*. *Applied and Environmental Microbiology*, 1986, 52 (4), 700-5.

23. Doyle, M.P. *Campylobacter jejuni*, in *Foodborne Diseases*. Ed. Cliver D.O. London. Academic Press, 1990, 218-22.

24. Doyle M.P., Roman D.J. Growth and survival of *Campylobacter* fetus subsp. jejuni as a function of temperature and pH. *Journal of Food Protection*, 1981, 44 (8), 596-601.

25. Stern N.J., Kazmi S.U. *Campylobacter jejuni*. Foodborne Bacterial Pathogens. Ed. Doyle M.P. New York. Marcel Dekker, 1989, 71-110.

26. Boucher S.N., Chamberlain A.H.L., Adfams M.R. Enhanced survival of *Campylobacter jejuni* in association with wood. *Journal of Food Protection,* 1998 (January), 61 (1), 26-30.

27. Patterson, M.F. Sensitivity of *Campylobacter* spp. to irradiation in poultry meat. *Letters in Applied Microbiology,* 1995, 20 (6), 338-40.

28. International Commission on Microbiological Specifications for Foods. *Campylobacter*, in *Microorganisms in foods, volume 5: microbiological specifications of food pathogens*. International Commission on Microbiological Specifications for Foods. London. Blackie, 1996, 45-65.

Further reading

Cramer M.M. Microorganisms of concern for food manufacturing in *Food plant sanitation: design, maintenance and good manufacturing practices*. Ed. Cramer M.M. Boca Ranton. CRC Press, 2006, 53-74.

Food Standards Agency. Control of campylobacters in extensively reared chickens: an investigation of growth inhibition and inactivation of campylobacters by plant extracts, in *FSA News Research Supplement*. London. FSA, 2006.

Havelaar A., Wagenaar J., Jacobs-Reitsma W. CARMA controls campylobacteriosis in the Netherlands. *World Poultry*, 2006, 22 (4), 41-43.

Altekruse S.F., Perez-Perez G.I. *Campylobacter jejuni* and related pathogens, in *Foodborne infections and intoxications*. Eds Riemann H.P., Cliver D.O. London. Academic Press, 2005, 259-87.

Linden J. *Campylobacter* gradually reveals its secrets. *Poultry International*, 2005, (December), 44 (13), 14-22.

McClure P., Blackburn C. *Campylobacter* and *Arcobacter*, in *Foodborne Pathogens: Hazards, Risk Analysis and Control.* Eds Blackburn C. de W., McClure P.J. Cambridge. Woodhead Publishing Ltd., 2002, 363-84.

Park S.F. The physiology of *Campylobacter* species and its relevance to their role as foodborne agents. *International Journal of Food Microbiology*, 2002, 77 (3), 177-88.

Various authors. *Campylobacter*, *Helicobacter* and *Arcobacter*. *Journal of Applied Microbiology*, 2001, 90, supplement 1-120.

Nachamkin, I. *Campylobacter jejuni*, in *Food Microbiology: Fundamentals and Frontiers.* 2nd edition. Eds Doyle M.P., Beuchat L.R., Montville T.J. Washington DC. ASM Press, 2001, 179-92.

Stern N.J., Line J.E. *Campylobacter.* The microbiological safety and quality of food, volume 2. Eds Lund B.M., Baird-Parker T.C., Gould G.W. Gaithersburg. Aspen Publishers, 2000, 1040-56.

Jay J.M. Foodborne gastroenteritis caused by *Vibrio*, *Yersinia*, and *Campylobacter* species, in *Modern food microbiology. 6th Edition.* Ed. Jay J.M. Gaithersburg. Aspen Publishers, 2000, 549-68.

Friedman C.R., Neimann J., Wegener H.C., Tauxe R.V. Epidemiology of *Campylobacter jejuni* infections in the United States and other industrialized nations, in *Campylobacter* (2nd Edition) Eds Nachamkin I., Blaser M.J. Washington DC. ASM Press, 2000, 121-38.

Methods of detection

de Boer E. Detection and enumeration of pathogens in meat, poultry and egg products, in *Microbiological analysis of red meat, poultry and eggs*. Ed. Mead G.C. Cambridge. Woodhead Publishing Ltd., 2007, 202-45.

Price E.P., Huygens F., Giffard P.M. Fingerprinting of *Campylobacter jejuni* by using resolution-optimized binary gene targets derived from comparative genome hybridization studies. *Applied and Environmental Microbiology*, 2006 (December), 72 (12), 7793-7803.

Fosse J., Laroche M., Rossero A., Federighi M., Seegers H., Magras C. Recovery methods for detection and quantification of *Campylobacter* depend on meat matrices and bacteriological or PCR tools. *Journal of Food Protection*, 2006 (September), 69 (9), 2100-06.

Hiett K.L., Seal B.S., Siragusa G.R. *Campylobacter* spp. subtype analysis using gel-based repetitive extragenic palindromic-PCR discriminates in parallel fashion to flaA short variable region DNA sequence analysis. *Journal of Applied Microbiology*, 2006 (December), 101 (6), 1249-58.

Zhang H., Gong Z., PuiO., Liu Y., Li X. F. An electronic DNA microarray technique for detection and differentiation of viable *Campylobacter* species. *Analyst*, 2006 (August), 131 (8), 907-15.

Barua R., Rathore R.S. Development of modified selective media for the isolation of *Campylobacter jejuni* from poultry. *Journal of Food Science and Technology*, 2006 (May-June), 43 (3), 305-7.

British Standards Institution. Microbiology of food and animal feeding stuffs. Horizontal method for detection and enumeration of *Campylobacter* spp. Detection method. BS EN ISO 10272-1, 2006.

Vollenhofer-Schrumpf S., Buresch R., Unger G., Stahl N., Franzl G., Schinkinger M. Detection of *Salmonella* spp., *Escherichia coli* O157, *Listeria monocytogenes* and *Campylobacter* spp. in chicken samples by multiplex polymerase chain reaction and hybridization using the GeneGen Major Food Pathogens Detection Kit. *Journal of Rapid Methods and Automation in Microbiology*, 2005 (September), 13 (3), 148-76.

Stern N.J., Line J.E., Chen H.-C. *Campylobacter* Compendium of Methods for the Microbiological Examination of Foods, 4th edn. Eds American Public Health Association, Downes F.P., Ito K. Washington DC. APHA, 2001, 301-10.

Wassenaar T.M., Newell D.G. Genotyping of *Campylobacter* spp. *Applied and Environmental Microbiology,* 2000, 66(1), 1-9.

Corry, J.E.L. Methods for the isolation of campylobacters and arcobacters, in *Factors Affecting the Microbial Quality of Meat, Volume 4: Microbial Methods for the Meat Industry.* Eds Commission of the European Communities, Hinton M.H., Rowlings C. Bristol. University of Bristol Press, 1996, 1.14.

Humphrey T.J., Martin K.W., Mason M.J. Isolation of *Campylobacter* species from non-clinical samples. *PHLS Microbiology Digest*, 1996, 13 (2), 86-88.

Hunt J.M., Abeyta C. *Campylobacter* Bacteriological Analytical Manual, 8th edition. Ed. Food and Drug Administration.. Gaithersburg. AOAC International, 1995, 27.

Food and Agriculture Organization, Andrews W. *Campylobacter* Manual of Food Quality Control, Vol. 4: Microbiological Analysis. Eds Food and Agriculture Organization, Andrews W. Rome. FAO, 1992, 87-108.

CLOSTRIDIUM BOTULINUM

The organism *Clostridium botulinum* was named after early investigations that linked outbreaks in central Europe with sausages. Hence the name 'botulism' from the Latin word 'botulus' meaning 'sausage'. In 1897, a Belgian physician found that an outbreak of alimentary intoxication was related to the 'ham bacillus' and called it *Bacillus botulinus*. However it was redesignated as *Clostridium botulinum* due to the phylogenetic difference between *Clostridium* and *Bacillus*.

The Organism *Clostridium botulinum*

Clostridia belong to the family Bacillaceae. *C. botulinum* is a Gram-positive, catalase and oxidase negative, spore-forming, motile rod with subterminal oval endospores and peritrichous flagellae. It can form toxins, and can grow only under anaerobic conditions.

There are four forms of botulism: 'classic' botulism, caused by ingestion of toxin in food; infant botulism (or 'floppy baby syndrome'); the rare wound botulism; and unclassified - this includes cases of unknown origin and adult cases that resemble infant botulism (1,2). This chapter considers foodborne and infant botulism.

Seven different types of *C. botulinum* are known, forming at least seven different toxins, A to G. Types A, B, E and, to a lesser extent, F are the types that are responsible for most cases of human botulism (2,3).

Botulism - the illness caused by the consumption of preformed botulinum toxin of *C. botulinum* - is extremely serious and, unless recognised and treated promptly, carries a high risk of mortality. It is, therefore, the most severe form of food poisoning.

The *C. botulinum* types can be divided into four groups based on their physiological differences (4).

Group	Types	Physiology
I	A,B,F	Proteolytic, mesophiles
II	B,E,F	Non-proteolytic, psychrotrophs
III*	C,D	Non-proteolytic
IV†	G	Weakly proteolytic

* these strains are not involved in human botulism
† no outbreaks of type G strains have been reported

Until recently *C. botulinum* was defined by the production of neurotoxin causing botulism, and this resulted in a very diverse species. Strains of other species have also been found to produce botulism neurotoxins (*C. butyricum* and *C. baratii*) and there have been reported outbreaks of foodborne illness associated with these organisms (5,6). It has been suggested that the four groups above should each be considered to be a separate species (7).

C. botulinum Food Poisoning

Clostridium botulinum was first linked with an outbreak of botulism in Belgium in 1895. Since then, a number of outbreaks have occurred all over the world, associated mainly with home-preserved vegetables, fish or meat products. The illness is most commonly associated with home-produced foods, or small catering units, rather than commercially prepared foods (3). Nevertheless, the impact of botulism associated with commercially manufactured foods, although rare, can be devastating.

'Classic botulism' is caused by ingestion of toxin that is pre-formed in food. *Clostridium botulinum* spores present in food must be given favourable conditions to germinate and grow to sufficient numbers to form toxin in the food, and the contaminated food must then be eaten raw, or without adequate heat treatment.

Incubation time

The incubation period is usually 12-36 h, but can be from a few hours up to 14 days, depending on the amount of toxin consumed. The earlier the symptoms occur, the more serious they are likely to be (2).

Symptoms

The severity of symptoms relates to the amount of toxin ingested, but the first systems are generally nausea and vomiting, followed by neurological signs and systems, including blurred vision and/or double vision, fixed and dilated pupils, loss of normal mouth and throat functions (dryness of the mouth, followed by difficulty in talking, swallowing, a sore throat), general fatigue and lack of muscle coordination, and respiratory impairment. Other gastrointestinal systems may include abdominal pain, diarrhoea, or constipation. Nausea and vomiting appear more often in cases of botulism types B and E; dysphagia and muscle weakness are more common in outbreaks of types A and B. (2). Very early vomiting often occurs with type E botulism, and this may be linked to lower mortality rates.

Mortality

Previously, the mortality rate for botulism was high, between 30 and 65%, with the rate being generally lower in European countries than in the US. Nowadays, the mortality rate is lower, below 10%, as a result of prompt treatment, administration of antitoxin and respiratory support. For example, the first outbreak to occur in the UK in 1922, which was caused by a commercially prepared duck paste, containing type A toxin, reported 8 cases and 8 deaths. In another UK outbreak in 1989 involving hazelnut yoghurt, 26 out of 27 people affected were admitted to hospital and only 1 person (an elderly woman) died (8).

In 1978, a single can of salmon imported from the USA, containing type E toxin, caused a family outbreak resulting in 4 cases and 2 deaths.

Infective dose

No live organisms of *C. botulinum* need to be consumed for foodborne botulism to occur. *C. botulinum* toxin is lethal - in fact, probably the most potent natural toxin known to man. It is believed that the oral lethal dose is between 0.005 and 0.1 μ for proteolytic types and between 0.1 and 0.5 μ for non-proteolytic types. As little as 0.1 g of food in which *C. botulinum* has grown and produced toxin can cause botulism (3). In effect, the presence of toxin at any concentration cannot be tolerated, and neither can conditions in foods in which toxin could be formed.

Foods involved

Foods involved in most cases and outbreaks of botulism have been associated with meat, fish and vegetables. In the USA (apart from Alaska), home-preserved (particularly canned) vegetables are the most frequent cause of classic botulism (commonly involving type A). In central Europe, meats, such as home-cured hams, are the foods most commonly responsible (particularly involving type B). In Italy and Spain, vegetables have been most commonly implicated (again involving type B). In northern Europe, Alaska, Canada, Scandinavia and Japan, fish products (containing type E) have been involved in the highest number of outbreaks (1,3,9).

The different patterns of botulism throughout the world reflect different eating habits and preservation methods. In the UK, where botulism is very rare, 'high risk' (including home-bottled/canned) foods are not normally eaten, and commercial canning and bottling processes are designed to ensure the destruction of spores of the organism.

The largest outbreak to occur in the UK was caused in 1989 by hazelnut conserve, which contained type B toxin, used in flavouring yoghurt. The conserve was inappropriately processed, following a change in formulation, allowing the survival and subsequent growth and toxin formation by *C. botulinum* in the canned product, which was later added to yoghurt (3,8).

Although most incidents of botulism have involved under-preserved meats, vegetables, or fish, more recent incidents worldwide have involved more unusual food types, including foil-wrapped baked potatoes used in potato salad, sautéed onions, cheese containing onion, Brie cheese, and temperature-abused garlic stored in oil (1,3,4,10,11).

Incidence of *C. botulinum* Food Poisoning

In the USA, an average of 9.4 outbreaks of botulism occur per year. Between 1950 and 1996 there were 44 foodborne outbreaks reported. The number of outbreaks per year increased up to 15-16 between 1978 and 1983, with an average of *ca* 3 cases/outbreak, and 17 deaths. During 1983-90 there were 107 outbreaks with 185 cases and 11 deaths. Therefore, the number of outbreaks, cases and deaths is decreasing with time (4). Currently, there are, on average, 110 cases of botulism reported in the US per year, of which 25% are foodborne and 72% are infant botulism. These figures have not changed significantly in recent years, but there has been an increase in wound botulism mainly associated with injecting drug users (source: CDC).

In the European Union, during the period 1988-1998, those countries reporting the highest numbers of cases of botulism were France (88 outbreaks of one or more cases), Germany (177), Italy (412) and Spain (92). Belgium, Denmark, England & Wales and Sweden generally reported less than one outbreak a year during the period 1988-1998 and outbreaks in these countries were usually small (less than three cases), with the exception of the hazelnut yoghurt outbreak in the UK in 1989 which resulted in 27 cases. (12, source: WHO).

Poland has the highest incidence of botulism in Europe. During 1988-1998 there were 1,995 outbreaks linked to the consumption of home-preserved meats (2). A recent paper reported nearly 2,000 cases with at least 28 deaths from 1988 to 1998 (13). In 1990, 74% of the 285 cases registered were linked to the consumption of home-processed foods, the other 26% were associated with commercially processed products, mainly canned meat and fish. (13).

In Japan, from 1988-1998, there were 51 cases of foodborne botulism and many of those cases were associated with izushi (fermented raw fish and cooked rice) (source: National Institute of Health Sciences, Japan).

Sources

Humans

Carriage of *C. botulinum* by humans does not play a significant role in the transmission of the organism.

Animals and environment

C. botulinum is ubiquitous - it is widely distributed in soil and marine sediments throughout the world. It is also found in the intestinal tract of animals, including fish. Surveys have shown geographical variations in the distribution of the different types of *C. botulinum* in the environment, particularly in soils. Type A predominates in the Western US, South America and China, whereas type B is more common in the Eastern US, the UK and Europe. Type E is more often found in Northern regions and in temperate aquatic environments. A recent survey of marine sediments in the Baltic sea showed that type E was present in 81% of samples (14).

Types A or B have occasionally been isolated from cattle, sheep and pigs, and have been isolated in very low numbers (generally 0.04 to 2.2 spore/kg) from meats including pork, bacon and liver sausage (15).

Foods

Because of the widespread occurrence of *C. botulinum* in the environment, the organism can be isolated occasionally from many foods. Most studies into the occurrence of the organism have concentrated on fish, meats and infant foods, especially honey. The highest incidence is in fish, such as farmed trout, Pacific salmon, and Baltic herring, which commonly contain low numbers of *C. botulinum*. Type E is the most commonly isolated type in fish, but other types may also be encountered (15).

Meat and meat products have been studied less intensively because there is considerably less contamination of farm environment (2).

Types A and B of Group I are commonly found in soil, and these organisms (especially type A) have been isolated from certain vegetables and fruit, including mushrooms. Other products in which contamination has often been detected include asparagus, beans, cabbage, carrots, celery, corn, onions, potatoes, turnips, olives, apricots, cherries, peaches and tomatoes (2). The organism has also been isolated from honey (55-60 spores/g) and other infant foods, namely corn syrup and rice cereal (but not in the UK) (2,15).

Growth/Survival Characteristics of the Organism in Foods

The prevention of growth and toxin production by *C. botulinum* - particularly by ensuring the destruction of the most heat-resistant spores of *C. botulinum* - is the basis of all commercial canning and many other food-preservation processes.

It is important to note that toxin production by *C. botulinum* does not necessarily coincide with obvious food spoilage. A food that contains toxin can be free of any odour, flavour or appearance that might warn the consumer that it is hazardous.

Temperature

All strains of *C. botulinum* grow reasonably well in the temperature range between 20 °C and 45 °C, but the low temperatures required to inhibit Groups I and II are different. Group I has an optimal temperature for growth at 37 °C with growth occurring between 10 °C and 48 °C (they will not grow at temperatures less than 10 °C), but Group II optimum growth temperature is between 25 °C and 30 °C. These strains are psychrotrophic, being capable of slow growth and toxin production (within 36 days) at low temperatures with an accepted minimum growth temperature of 3 °C. This ability to grow at chill temperatures is of particular concern in relation to products that are given a mild heat treatment under, or followed by, modified-atmosphere/vacuum packaging, and are expected to have a relatively long shelf-life at chill temperatures (16).

Continuous storage of products at <3.3 °C could be used as a method for the control of growth and toxin production. However, in practice, the temperatures used for chill food storage, by retailers and consumers, may allow growth and toxin production, and thus this is one of the major contributory factors in most foodborne botulism outbreaks in the US. It should be noted that, in a study into the incidence of psychrotrophic *C. botulinum* in refrigerated, packaged (vacuum, overwrap or modified atmosphere) foods in the UK, the organism was not detected (17).

Heat resistance

While the vegetative cells of *C. botulinum* are not particularly heat-resistant, the spores of this organism are more so. Spores of Group I (proteolytic) strains are those with the greatest heat resistance. Thermal inactivation times (D-values) vary between the *C. botulinum* strains, and are dependent on the way in which the spores were produced and treated, the heating environment, and the recovery system. The most heat-resistant spores of Group I *C. botulinum* are produced by type A and proteolytic B strains ($D_{121\,°C}$ = *ca* 0.21 min) (18).

The heat resistance of these Group I spores of *C. botulinum* has been studied extensively, as it is the basis of safe canning processes for foods that have a pH value above 4.5 ('low-acid'). A 'Botulinum Cook' is required for foods destined to be stored at ambient temperatures (principally canned foods) that are capable of supporting the growth of *C. botulinum*. These foods are those with a pH value >4.5 and a_w that is high enough to allow growth. In theory this is >0.93, but a margin for safety is recommended; in the USA, low-acid foods are defined as those with a pH greater than pH 4.6.

A Botulinum Cook is defined as that giving a 12 log cycle kill, i.e. destruction of 10^{12} spores of the most resistant proteolytic strain of *C. botulinum*. In commercial terms, this is a cook of equivalent to at least 3 min at 121 °C at the slowest heating point in the container, which then allows the safe storage of foods at ambient temperatures (4).

The spores of Group II (non-proteolytic/psychrotrophic) strains are less heat-resistant than those of Group I strains. However, they may survive processing at temperatures of 70-85 °C and their ability to grow at refrigeration temperatures necessitates their control in foods capable of supporting their growth (e.g. vacuum-packed, par-cooked meals with pH value >5.0 and a_w >0.97) (19,20).

All *C. botulinum* types produce heat labile toxins, which may be inactivated by heating at 80 °C for 20-30 min, at 85 °C for 5 min, or at 90 °C for a few seconds.

The potential for growth of psychrotrophic *C. botulinum* in REPFEDS (refrigerated processed foods with extended durability) or 'sous vide' products has led to a number of authors and associations recommending appropriate processing conditions. In general, it is recommended that the product is heated at 90 °C for 10 min to achieve a 6 log cycle kill. For some products, it may not be appropriate to follow these process conditions. In these cases, a combination of heat and other factors should be used to prevent the growth of *C. botulinum*, e.g. a_w <0.97, pH <5.0, and salt >3.5%.

The control of psychrotrophic strains of *C. botulinum* in chilled, reduced-oxygen-packed food is considered in detail in the 'Report on Vacuum Packaging and Associated Processes', by the Advisory Committee on the Microbiological Safety of Food (21). Food producers are strongly advised to follow the advice given in this report in establishing appropriate heat treatments and shelf-lives for any vacuum- or modified-atmosphere-packed products that might have the potential to support the growth of *C. botulinum*.

pH

Generally, it is accepted that the minimum pH for the growth of *C. botulinum* Group I is pH 4.6, and many regulations worldwide use this limit. The limit for Group II organisms is pH 5.0 (1,18). However, there have been examples of growth and toxin production at pH levels below 4.6. These data were obtained in laboratory media, or under laboratory conditions, and the presence of high concentrations of proteins protected the organism (16). This does not occur in foods preserved by acidity (18).

As pH has a major bearing on the ability of foods to support the growth of *C. botulinum*, foods are traditionally categorised into 'high-acid' and 'low-acid' products. In the UK, low-acid products are defined as those having a pH value equal to, or greater than, 4.5. These products must be given a Botulinum Cook or be controlled by some other factor such as water activity. (Note that in the Department of Health guidelines for the canning of low-acid foods, if a heat process less than the Botulinum Cook is given, the onus is on the canner to show that the process provides equivalent protection against *C. botulinum*) (22).

The growth of acid-tolerant micro-organisms, such as yeasts, moulds and bacilli can increase the pH in their immediate area to a level that allows growth of *C. botulinum*. All of the 35 outbreaks of botulism attributed to normally 'high-acid' foods had been spoiled by yeasts or moulds (23).

Water activity/salt

Water activity (a_w) in foods can be altered by several different solutes. The most common humectant used in foods is sodium chloride (NaCl) or salt. It is generally accepted that under optimal growth conditions 10% (w/w) NaCl is required to prevent growth of Group I organisms and 5% (w/w) NaCl is necessary to prevent growth of Group II organisms. These concentrations correspond to limiting a_w of 0.94 for Group I and 0.97 for Group II (2). The type of solute used to control a_w may influence these minimum values. Generally, sodium chloride, potassium chloride, glucose and sucrose result in similar limits; however, glycerol will allow growth at lower a_w values (by up to 0.03 units) (16,18).

These values have been established under carefully controlled laboratory conditions. In commercial situations, safety margins must be introduced to allow for process variability.

Atmosphere

Although *C. botulinum* is a strict anaerobe, many foods that are not obviously 'anaerobic' can provide adequate conditions for growth. Thus, an aerobically packed product may not support the growth of the organism on the surface, but the interior of the food may do so. It is also important to note that the inclusion of oxygen (O_2) as a packaging gas cannot ensure that growth of *C. botulinum* is prevented; the Eh (reduction-oxidation potential or redox) of most such foods is usually low enough to permit its growth (2,16). In an investigation of *C. botulinum* toxin production in inoculated pork, it was shown that initial atmospheres with 20% oxygen (O_2) did not delay toxin production, compared with samples packaged in 100% nitrogen (N_2) (1,2).

Increasingly, modified-atmosphere packaging (MAP) is being used to extend the shelf-life, and improve the quality of foods. However, depending on the particular atmosphere, it may allow the growth of *C. botulinum*. Many MAP products utilise carbon dioxide (CO_2) to control spoilage and pathogenic bacteria, but CO_2 may allow the growth of *C. botulinum* dependent upon the gas concentration. Studies with inoculated pork showed that CO_2 levels of 15-30% did not inhibit *C. botulinum*; only 75% CO_2 showed significant inhibition (1,2,16).

Summary of factors controlling the growth of* C. botulinum *in foods (6)

	Group I (proteolytic)	**Group II (non-proteolytic)**
	A,B,F	B,E,F
pH	<4.6	<5.0
a_w	<0.94	<0.97
Inhibitory (NaCl)	10%	5%
Temperature	<10 °C	<3 °C
$D_{100\,°C}$ of spores	25 min	0.1 min
$D_{121\,°C}$ of spores	0.1-0.2 min	<0.001 min

Summary of Control of *C. botulinum* in Foods

It must be assumed that spores of *C. botulinum* may be present in raw foods. Processing or preservation conditions must, therefore, be adequate to destroy spores or to prevent germination, subsequent growth and toxin production. This relies on the appropriate use of heat, manipulation of pH, water activity, addition of nitrite or other preservatives, temperature control, etc. Appropriate codes of practice should be followed where available. Changes in established, validated formulations or processes, whether intentional or inadvertent, must be evaluated by a hazard analysis, and appropriate controls instituted for those parameters responsible for destroying or controlling the growth of the organism.

A variety of chemical preservatives have been shown to inhibit *C. botulinum*, either alone or in combination. These include: nitrites, sorbates, parabens, nisin, phenolic antioxidants, polyphosphates, ascorbates, EDTA, metabisulfite, n-monoakyl maleates and fumarates, and lactate salts. In fish, the addition of liquid smoke has been found to reduce the amount of salt needed to inhibit growth. Comprehensive reviews on the effects of these preservatives have been published (1,7,16,18,24).

Note: Infant Botulism

Infant botulism was first reported in 1976, and since that time over 1,000 cases have been confirmed worldwide. In the USA, infant botulism is now the most common form of botulism (1,2). However, it appears to be very rare in Europe, where there have only been 49 recorded cases, of which 6 were in the UK (25).

Toxin-producing *C. butyricum* and *C. baratii* have also been implicated in 7 cases of 'infant botulism' (25).

It is thought that, in infants under a year old, ingested spores of *C. botulinum* (which do not pose a direct hazard to adults because of the presence of an inhibitory normal gut flora) can germinate and multiply in the intestine, forming toxin *in situ*. Symptoms include generalised weakness and weak cry. They may also include feeding difficulty and poor sucking, lethargy, lack of facial expression, irritability and progressive 'floppiness'

Honey (and possibly glucose syrup), that is sometimes used on 'pacifiers' for infants, may attract spore-contaminated dust. Contamination of the honey may result from growth of the organism in dead larvae present in bee hives. The prevalence of *C. botulinum* spores in honey is usually low (typically <1 cfu/g) but they are occasionally found at levels as high as 60 cfu/g. (26). 80 spores/g of *C. botulinum* types A and B were reported in one sample of honey incriminated in an incident of infant botulism (27), and the minimum infective dose of *C. botulinum* spores for infants has been estimated to be as low as 10-100 spores (25). Although many studies have failed to detect *C. botulinum* spores in honey, 5.1 % samples, out of 2,033 tested, from various parts of the world contained detectable levels of *C. botulinum* spores (26).

Bibliography

References

1. Dodds K.L. *Clostridium botulinum*, in *Foodborne Disease Handbook, Vol. 1: Diseases Caused by Bacteria*. Eds Hui Y.H., Gorman J.R., Murrell K.D., Cliver D.O. New York. Marcel Dekker, 1994, 97-131.

2. Austin J. *Clostridium botulinum,* in *Food Microbiology: Fundamentals and Frontiers*. Eds Doyle M.P., Beuchat L.R., Montville T.J. Washington DC. ASM Press, 2001, 329-49.

3. Novak J., Peck M., Juneja V., Johnson E. *Clostridium botulinum* and *Clostridium perfringens,* in *Foodborne Pathogens. Microbiology and Molecular Biology*. Fratamico P.M., Bhunia A.K., Smith J.L. Great Britain. Caister Academic Press, 2005, 383-408.

4. Rhodehamel E.J., Reddy N.R., Pierson M.D. Botulism: the causative agent and its control in foods. *Fd Control,* 1992, 3 (3), 125-43

5. Anniballi F., Fenicia L., Franciosa G., Aureli P. Influence of pH and temperature on the growth of and toxin production by neurotoxigenic strains of *Clostridium butyricum* type E. *Journal of Food Protection,* 2002, 65 (8), 1267-70.

6. Harvey S.M., Sturgeon J., Dassey D.E. Botulism due to *Clostridium baratii* type F toxin. *Journal of Clinical Microbiology,* 2002, 40 (6), 2260-2.

7. Collins M.D., East A.K. A review: phylogeny and taxonomy of the foodborne pathogen *Clostridium botulinum* and its neurotoxins. *Journal of Applied Microbiology,* 1998, (January), 84 (1), 5-17.

8. O'Mahony M., Mitchell E., Gilbert R.J., Hutchinson D.N., Begg N.T., Rodhouse J.C., Morris J.E. An outbreak of foodborne botulism associated with contaminated hazelnut yoghurt. *Epidemiology and Infection,* 1990, 104 (3), 389-95.

9. Hauschild A.H.W. Epidemiology of human foodborne botulism. *Clostridium botulinum*: Ecology and Control in Foods. Eds Hauschild A.H.W., Dodds K.L. New York. Marcel Dekker, 1993, 69-104.

10. St. Louis M.E., Peck S.H.S., Bowering D. Botulism from chopped garlic: delayed recognition of a major outbreak. *Am. Intern. Med.*, 1988, 108 (3), 363-8.

11. Solomon H.G., Kautter D.A. Growth and toxin production by *Clostridium botulinum* in sautéed onions. *Journal of Food Protection,* 1986, 49 (8), 618-20.

12. Therre H. Botulism in the European Union. *Eurosurveillance Monthly,* 1999 (January), 4 (1), 2-7.

13. Galazka A., Przybylska A. Surveillance of foodborne botulism in Poland: 1960-1998 Eurosurveillance Monthly 1999 (June) 4 (6), 69-72.

14. Hielm S., Hyytia E., Andersin A.-B., Korkeala H. A high prevalence of *Clostridium botulinum* type E in Finnish freshwater and Baltic Sea sediment samples. *Journal of Applied Microbiology,* 1998, (January), 84 (1), 133-7.

15. Franciosa G., Aureli P., Schecter R. *Clostridium botulinum* in International Handbook of Foodborne Pathogens. Eds Miliotis M.D., Bier J.W. New York. Marcel Dekker Inc., 2003, 61-90.

16. Kim J., Foegeding P.M. Principles of control *Clostridium botulinum*: Econology and Control in Foods. Eds Hauschild A.H.W., Dodds K.L. New York. Marcel Dekker, 1993 121-76.

17. Gibbs P.A., Davies A.R., Fletcher R.S. Incidence and growth of psychrotrophic *Clostridium botulinum* in foods. *Food Control,* 1992, 3 (3), 125-43.

18. Hauschild A.H.W. *Clostridium botulinum*, in *Foodborne Bacterial Pathogens.* Ed. Doyle M.P. New York. Marcel Dekker, 1989, 111-89.

19. Lund B.M., Notermans S.H.W. Potential hazards associated with REPFEDS *Clostridium botulinum*: Ecology and Control in Foods. Eds Hauschild A.H.W., Dodds K.L. New York. Marcel Dekker, 1993, 279-303.

20. Betts G.D., Gaze J.E. Growth and heat resistance of psychrotrophic *Clostridium botulinum* in relation to 'sous vide' *Food Control,* 1995, 6 (1), 57-63.

21. Advisory Committee on the Microbiological Safety of Food. Report on vacuum packaging and associated processes. London: HMSO 1992.

22. Department of Health and Social Security, Ministry of Agriculture Fisheries and Food, Scottish Home and Health Department, Department of Health and Social Services Northern Ireland, Welsh Office. Food hygiene codes of practice no. 10. The canning of low acid foods: a guide to good manufacturing practice. London. HMSO, 1981.

23. Solomon H.M., Kautter D.A. Outgrowth and toxin production by *Clostridium botulinum* in bottled chopped garlic. *Journal of Food Protection,* 1988, 51 (11), 862-5.

24. Szabo E.A., Gibson A.M. *Clostridium botulinum,* in *Foodborne Microorganisms of Public Health Significance.* Australian Institute of Food Science and Technology 5th Edn. North Sydney: AIFST, 1997 429-64.

25. Scientific Committee on Veterinary Measures Relating to Public Health. Opinion of the Scientific Committee on Veterinary Measures Relating to Public Health on honey and microbiological hazards. Brussels. European Commission, 2002.

26. Snowdon J.A., Cliver D.O. Microorganisms in honey. *International Journal of Food Microbiology,* 1996, 31 (1-3), 1-26.

27. Dodds K.L. Worldwide incidence and ecology of infant botulism *Clostridium botulinum*: Ecology and Control in Foods. Eds Hauschild A.H.W., Dodds K.L. New York. Marcel Dekker, 1993, 105-17.

Further Reading

Advisory Committee on the Microbiological Safety of Food, Food Standards Agency. Report on botulism in cattle by the Ad Hoc Group on Botulism in Cattle. London. FSA, 2006, 68.

Austin J.W., Smith J.P. Botulism from fishery products: history and control, in *Modified atmosphere processing and packaging of fish: filtered smokes, carbon monoxide, and reduced oxygen packaging.* Eds Otwell W.S., Balban M.O., Kristinsson H.G. Oxford. Blackwell Publishing, 2006, 93-216.

Skinner G.E., Reddy N.R. Hazards associated with *Clostridium botulinum* in modified atmosphere packaged fresh fish and fishery products, in *Modified atmosphere processing and packaging of fish: filtered smokes, carbon monoxide, and reduced oxygen packaging.* Eds Otwell W.S., Balban M.O., Kristinsson H.G. Oxford. Blackwell Publishing. 2006, 163-92.

Advisory Committee on the Microbiological Safety of Food, Food Standards Agency. Report on minimally processed infant weaning foods and the risk of infant botulism. London. FSA, 2006, 54.

Lindstrom M., Kiviniemi K., Korkeala H. Hazard and control of group II (non-proteolytic) *Clostridium botulinum* in modern food processing. *International Journal of Food Microbiology*, 2006, (April 15), 108 (1), 92-104.

Montville T.J., Matthews K.R. *Clostridium botulinum,* in *Food microbiology: an introduction.* Eds Montville T.J., Matthews K.R. Washington DC. ASM Press, 2005, 187-99.

Bell C., Kriakides A. *Clostridium botulinum*: a practical approach to the organism and its control in foods. Oxford. Blackwell Science, 2000.

Lund B.M., Peck M.W. *Clostridium botulinum*. The microbiological safety and quality of food, volume 2. Eds Lund B.M., Baird-Parker T.C., Gould G.W. Gaithersburg. Aspen Publishers, 2000, 1057-1109.

Juneja V.K. Hazards associated with non-proteolytic *Clostridium botulinum* and other spore-formers in extended-life refrigerated foods. Sous vide and Cook-chill Processing for the Food Industry. Ed. Ghazala S. Gaithersburg. Aspen Publishers, 1998, 234-73.

Szabo E.A., Gibson A.M. *Clostridium botulinum*, in *Foodborne Microorganisms of Public Health Significance.* Ed. Australian Institute of Food Science and Technology. 5th Edn. North Sydney. AIFST, 1997, 429-64.

Peck M.W. *Clostridium botulinum* and the safety of refrigerated processed foods of extended durability. *Trends in Food Science and Technology,* 1997 (June), 8 (6), 186-92

Peck M.W. *Clostridium botulinum* and processed meat products. Factors Affecting the Microbial Quality of Meat, Volume 3: Cutting and Further Processing. Eds Commission of the European Communities, Hinton M.H., Rowlings C. Bristol. University of Bristol Press, 1996, 165-8.

International Commission on Microbiological Specifications for Foods. *Clostridium botulinum*, in *Microorganisms in Foods, Volume 5: Microbiological Specifications of Food Pathogens.* Ed. International Commissionon Microbiological Specifications for Foods. London. Blackie, 1996, 66-111.

Methods of detection

Kennett C.A., Stark B. Automated ribotyping for the identification and characterization of foodborne Clostridia. *Journal of Food Protection,* 2006 (December), 69 (12), 2970-5.

Kirkwood J., Ghetler A., Sedman J., Leclair D., Pagotto F., Austin J.W., Ismail A.A. Differentiation of group I and group II strains of *Clostridium botulinum* by focal plane array Fourier transform infrared spectroscopy. *Journal of Food Protection*, 2006, (October), 69 (10), 2377-83.

Sharma S.K., Ferreirra J.L., Eblen B.S., Whiting R.C. Detection of type A, B, E and F *Clostridium botulinum* neurotoxins in foods by using an amplified enzyme-linked immunosorbent assay with digoxigenin-labeled antibodies. *Applied and Environmental Microbiology*, 2006 (February), 72 (2), 1231-8.

Lindstrom M., Nevas M., Korkeala H. Detection of *Clostridium botulinum* by multiplex PCR in foods and faeces in Food-borne pathogens: methods and protocols. Ed. Adley C.C. Totowa. Humana Press, 2005, 37-45.

Sharma S.K., Eblen B.S., Bull R.L., Burr D.H., Whiting R.C. Evaluation of lateral-flow *Clostridium botulinum* neurotoxin detection kits for food analysis. *Applied and Environmental Microbiology*, 2005 (July), 71 (7), 3935-41.

Gauthier M., Cadieux B., Austin J.W., Blais B.W. Cloth-based hybridization array system for the detection of *Clostridium botulinum* type A, B, E and F neurotoxin genes. *Journal of Food Protection*, 2005 (July), 68 (7), 1477-83.

Barr J.R., Moura H., Boyer A.E., Woolfitt A.R., Kalb S.R., Pavlopoulos A., McWilliams L.G., Schmidt J.G., Martinez R.A., Ashley D.L. Botulinum neurotoxin detection and differentiation by mass spectrometry. *Emerging Infectious Diseases,* 2005 (October), 11 (10), 11.

Akbulut D., Grant K.A., McLauchlin J. Development and application of real-time PCR assays to detect fragments of the *Clostridium botulinum* types A, B, and E neurotoxin genes for investigation of human foodborne and infant botulism. *Foodborne Pathogens and Disease.* 2004 (Winter), 1 (4), 247-257.

Solomon H.M., Johnson E.A., Bernard D.T., Arnon S.S., Ferreira J.L. *Clostridium botulinum* and its toxins, in *Compendium of Methods for the Microbiological Examination of Foods, 4th edition.* Eds American Public Health Association, Downes F.P., Ito K. Washington DC. APHA, 2001, 317-24.

Wictome M., Newton K., Jameson K., Dunnigan P., Clarke S., Wright S., Gaze J., Tauk A., Foster K.A., Shone C.C. Evaluation of novel *in vitro* assays for the detection of botulinum toxins in foods. Rapid detection assays for food and water. Proceedings of the International Conference on Developments in Rapid Diagnostic Methods: Water and Food, York, March 1999. Eds Clark S., Thompson K.C., Keevil C.W., Smith M. Cambridge. RSC, 2001, 206-9.

Cantoni C., Marchisio E., Butta P. Evaluation of some media for the isolation of *C. (Clostridium) botulinum. Industrie Alimentari,* 1998, 37 (366), 33-6.

Aranda E., Roderiguez M.M., Asenio M.A., Cordoba J.J. Detection of *Clostridium botulinum* types A, B, E and F in foods by PCR and DNA probe. *Letters in Applied Microbiology,* 1997 (September), 25 (3), 186-90.

Solomon H.M., Rhodehamel E.J., Kautter D.A. *Clostridium botulinum.* Bacteriological Analytical Manual, 8th Edn. Ed. Food and Drug Administration. Gaithersburg. AOAC International, 1995.

Food and Agriculture Organization, Andrews W. *Clostridium botulinum,* in *Manual of Food Quality Control, Volume 4: Microbiological Analysis*. Eds Food and Agriculture Organization, Andrews W. Rome. FAO, 1992, 213-20.

CLOSTRIDIUM PERFRINGENS

Clostridium perfringens was first described as *Bacillus aerogenes* in 1892 and later called *Clostridium welchii*. The main species of *Clostridium* that is of concern in the context of food safety is *Clostridium botulinum*. However, *Clostridium perfringens* is the cause of a much less severe but more common type of food poisoning.

The Organism *Clostridium perfringens*

Like other species of clostridia, *C. perfringens* is a Gram-positive, catalase and oxidase negative, non-motile, encapsulated bacillus belonging to the family Bacillaceae. Also, like other clostridia, it forms spores (endospores), which are heat-resistant, surviving normal cooking conditions. It is also anaerobic, but can tolerate moderate exposure to air.

Strains of *C. perfringens* are classified into five types, A-E, according to the toxin produced, and it is *C. perfringens* type A and, to a lesser extent, C, that is of primary concern in the context of food safety (1,2).

C. perfringens Food Poisoning

This organism first earned its reputation as a pathogen as the cause of gas gangrene, but its association with food poisoning has been recognised since the 1940s. It has since become recognised as a significant cause of food poisoning worldwide (3).

C. perfringens is capable of producing at least 13 different toxins, and it is primarily the formation of enterotoxin that is responsible for the symptoms of food poisoning by this organism. This toxin is not normally formed by *C. perfringens* until the organism enters the human intestine and begins to multiply. It is commonly thought that the organism then sporulates, at the same time releasing the enterotoxin responsible for the symptoms

(1,2). Symptoms are therefore the result of an infection rather than intoxication.

Incidents of rapid onset symptoms have indicated that a toxin may be preformed in foods. However, this would appear to be rare; the enterotoxin is not normally formed in sufficient quantities in food to cause illness, but preformed toxin may contribute to the rapid development of symptoms (3,4).

Incubation time

Symptoms of *C. perfringens* food poisoning occur between 8 and 22 h (usually 12-18 h) after the ingestion of contaminated food.

Symptoms

The symptoms of food poisoning caused by *C. perfringens* normally comprise diarrhoea and severe abdominal pain. Nausea occurs occasionally, but fever and vomiting are unusual. The illness is a relatively mild one (5).

Mortality

Recovery from *C. perfringens* is normally complete within 24 h. However, complications and death from *C. perfringens* have been recorded on very rare occasions, amongst elderly debilitated people (5).

Infective dose

Large numbers (>10^5/g - typically 10^6-10^8/g) of viable vegetative cells of *C. perfringens* are needed in foods to cause food poisoning by this organism (5,6).

Foods involved

Most outbreaks of food poisoning caused by *C. perfringens* are associated with meat dishes; especially beef, as well as poultry. Such food poisoning is most commonly reported as a consequence of institutional and restaurant or reception catering. Virtually no outbreaks have been linked with commercially prepared processed foods (1-4).

Spores of *C. perfringens* on meat or poultry can survive cooking (especially if protected within the cavity of poultry carcases, or in the centre of rolled roasts or casseroles, stews, pies, etc.), and may in fact be stimulated to germinate by the heat applied (6). Inadequate cooling facilities for cooked foods, leading to slow cooling, can allow germination and rapid multiplication of the organisms to numbers that constitute an infective dose. This happens most commonly during mass catering; if large quantities of food need to be cooked well in advance of being consumed, and facilities are not adequate for rapid chilling and refrigerated storage. Dishes containing high numbers of *C. perfringens* may remain palatable, giving no warning to the consumer of any hazard. Spores are not usually present (4,6,7).

Incidence of *C. perfringens* Food Poisoning

As is the case with other minor forms of food poisoning, it is thought that the vast majority of incidents of food poisoning caused by *C. perfringens* (perhaps 90-95%) go unreported. Nevertheless, *C. perfringens* features quite significantly amongst the food-poisoning statistics both in the UK and in the US (6).

In England and Wales, the reported incidence of food poisoning attributed to *C. perfringens*, although low in comparison with that caused by *Salmonella*, is greater than that of *B. cereus* and *Staph. aureus*; after *Salmonella* and *Campylobacter*, it is the most commonly reported agent. There was an average of 50 outbreaks of food poisoning per year reported to CDSC between 1989 and 1991 attributed to *C. perfringens*, totalling 3,064 cases (8).

During 1992-93, there were only 68 reported outbreaks due to *C. perfringens* - an average of 34 per year (9). Therefore, this type of food poisoning appears to be decreasing in incidence in the UK. The downward trend continued in 1994-1995, but, in 1996, 33 general outbreaks and 720 individual cases were recorded in England and Wales (source: CDSC).

Similar trends have been observed in the US; during 1983-87, there were 24 confirmed *C. perfringens* outbreaks, totalling 2,743 cases and 2 deaths (10). During 1993-97, 57 outbreaks and 2,772 cases of *C. perfringens* food poisoning were reported by the CDC (11). In 1999, it was estimated that there were approximately 250,000 annual cases of foodborne illness in the US caused by *C. perfringens*, resulting in 41 patients requiring hospitalisation and 7 deaths (12).

Sources

Humans

Clostridium perfringens spores are present in numbers between 10^4 and 10^6 as part of the normal faecal flora of most individuals.

The high incidence of *C. perfringens* in faecal flora makes it necessary to detect enterotoxin from isolates for outbreak confirmation (1).

Animals and environment

The spores of *C. perfringens* type A are widely distributed in the environment, and are usually present in soil - especially well-manured soil - in high numbers (10^3-10^4/g).

Spores of *C. perfringens* are commonly found in the faeces of many animals. The number of *C. perfringens* in the intestinal tract varies from animal to animal, and among individuals of the same species.

Foods

C. perfringens is commonly found in low numbers in many foods, especially in meat and poultry and their products, and in Mexican foods (3). Because of the common occurrence of *C. perfringens*, the interpretation of the significance of finding low numbers of *C. perfringens* in foods - especially meat and poultry and their products - should be made with caution; many wholesome foods may contain low numbers of the organism (13). It has been suggested that only strains of *C. perfringens* that have been subjected to repeated heating are able to cause food poisoning, and that strains freshly isolated from the environment do not (14).

Growth/Survival Characteristics of the Organism in Foods

Temperature

The most significant characteristic of *C. perfringens* in relation to food safety is the organism's ability to grow extremely rapidly at high temperatures. Its optimum temperature for growth is 43-45 °C. At these temperatures, *C. perfringens* has one of the fastest rates of growth (shortest generation times) of any bacterium; a generation time of approximately 7 min has been recorded at 41 °C for one strain of *C. perfringens*, although 10 min is more typical (3,6).

However, *C. perfringens* has the potential ability to grow within the temperature range 15-50 °C, depending on strain and other conditions. Although some growth can occur at 50 °C, death of the vegetative cells of this organism usually occurs rapidly above this temperature (1,3). At cold temperatures, 0-10 °C, vegetative cells die rapidly (1).

Heat resistance

Exposure to a temperature of 60 °C or more will result in the death of vegetative cells of *C. perfringens*, although prior growth at high temperatures or the presence of fat in a food will result in increased heat resistance. (It is unusual for spores to be formed in foods after the growth of this organism) (15).

The spores of *C. perfringens* can vary quite considerably in their heat resistance, and their heat resistance is also affected by the heating substrate. Recorded heat resistance values at 95 °C (D-values) range from 17.6-63 min for heat-resistant spores to 1.3-2.8 min for heat-sensitive spores (1). Recent studies suggest that at 55 °C, vegetative cells of food poisoning isolates are about twice as heat resistant as vegetative cells of other *C. perfringens* isolates (5).

In addition, the enterotoxin is not heat-resistant - it is destroyed by heating at 60 °C for 10 min (4,6,15).

pH

C. perfringens is not a tolerant organism with respect to pH. It grows best at pH values between 6 and 7 (the same pH as most meats). Under otherwise ideal conditions, very limited growth may occur at pH values over the range $\leq$5 and $\geq$8.3. Spores, however, will survive greater extremes of pH (and a_w) (1,3).

Water activity/salt

C. perfringens is not tolerant of low water activities. As in the case of other factors limiting the growth/survival of this organism, the limits for water activity are affected by temperature, pH, type of solute, etc. The lowest a_w recorded to support the growth of *C. perfringens* is 0.93 to 0.97 depending on the solute used to control the a_w of the medium (3,5).

Salt concentrations of 6-8% inhibit growth of most *C. perfringens* strains, but lower concentrations may be effective in combination with other factors. Some studies have indicated that the presence of 3% NaCl delays growth of *C. perfringens* in vacuum-packed beef (5).

Atmosphere

C. perfringens - like other clostridia - is an anaerobe. It will not, therefore, grow on the surface of foods unless they are vacuum-or gas-packed. The organism will grow well in the centre of meat or poultry dishes, where oxygen levels are reduced particularly by cooking (15).

The oxidation-reduction potential (Redox or Eh) for growth is reported to be between -125mV and +300mV (1,3). Once growth is initiated, the cells are able to modify the local Eh to favour more rapid growth, probably by production of substances such as ferredoxin (15).

Summary of Control of *C. perfringens* in Foods

Food poisoning from *C. perfringens* occurs most commonly where there is inadequate temperature control after cooking. Improper cooking of foods and contaminated equipment also contribute to infection. The prevention of food poisoning by *C. perfringens* can be assured by thoroughly cooking the food. In addition, prevention can be guaranteed by the rapid cooling of any large meat or poultry portions or dishes after cooking to temperatures below those allowing the growth of the organism (below 15 °C, or preferably below 10 °C, within 2-3 h). Alternatively, such cooked foods should be kept until serving at temperatures too high for the growth of *C. perfringens* (>60 °C).

Pre-cooked foods should be reheated before serving, to an internal temperature of at least 75 °C, to ensure the destruction of vegetative cells of *C. perfringens*.

Bibliography

References

1. Wrigley D.M. *Clostridium perfringens*, in *Foodborne Disease Handbook, Vol. 1. Diseases Caused by Bacteria.* Eds Hui Y.H., Gorham J.R., Murrell K.D., Cliver D.O. New York. Marcel Dekker, 1994, 133-67.

2. International Commission on Microbiological Specifications for Foods. *Clostridium perfringens*, in *Microorganisms in Foods, Vol. 5. Microbiological Specifications of Food Pathogens.* Ed. International Commission on Microbiological Specifications for Foods. London. Blackie, 1996, 112-25.

3. Labbe R., Juneja V.K. *Clostridium perfringens* gastroenteritis, in *Foodborne Infection and Intoxication.* Eds Riemann H.P., Cliver D.O. London. Elsevier, 2006, 137-64.

4. Lund B.M. Foodborne disease due to *Bacillus* and *Clostridium* species. *Lancet,* 1990, 336 (8721), 982-6.

5. McClane B.A. *Clostridium perfringens,* in *Food Microbiology: Fundamentals and Frontiers.* Eds Doyle M.P., Beuchat L.R., Montville T.J. Washington DC. ASM Press, 2001, 351-82.

6. Johnson E.A. *Clostridium perfringens* food poisoning. Foodborne Diseases. Ed. Cliver D.O. London. Academic Press, 1990, 229-40.

7. Reed G.H. Foodborne illness (Part 3). *Clostridium perfringens* gastroenteritis. *Dairy, Food and Environmental Sanitation,* 1994, 14 (1), 16-7.

8. Sockett P.N., Cowden J.M., Le Baigue S., Ross D., Adak G.K., Evans H. Foodborne disease surveillance in England and Wales: 1989-91. *CDR*, 1993, 3 (12), R159-73.

9. Cowden J.M., Wall P.G., Adak G., Evans H., Le Baigue S., Ross D. Outbreaks of foodborne infectious intestinal disease in England and Wales: 1992 and 1993. *CDR Review*, 1995, 5 (8), R109-17.

10. Jay J., Loessner M., Golden D. Food Poisoning Caused by Gram-Positive Sporeforming Bacteria, in *Modern Food Microbiology*. USA. Springer, 2005, 567-72.

11. Bean N.H., Griffin P.M., Goulding J.S., Ivey C.B. Foodborne disease outbreaks, 5-year summary, 1983-7. *Morbidity and Mortality Weekly Report,* 1990, 39 (SS-1), 15-57.

12. Mead P.S. *et al*. Food-related illness and death in the United States. *Emerging Infectious Disease*, 1999, 5 (5), 607-25.

13. International Commission on Microbiological Specifications for Food. Microorganisms in foods, Vol. 1. Their significance and methods of enumeration. Toronto. University of Toronto Press, 2nd rev. 1988, 436.

14. Andersson A., Ronner U., Granum P.E. What problems does the food industry have with the spore-forming pathogens *Bacillus cereus* and *Clostridium perfringens*? *International Journal of Food Microbiology,* 1995, 28 (2), 145-55.

15. Labbe R. *Clostridium perfringens,* in *Foodborne Bacterial Pathogens*. Eds Doyle M.P. New York. Marcel Dekker, 1989, 191-243.

Further reading

Li J., McClane B.A. Comparative effects of osmotic, sodium nitrite-induced, and pH-induced stress on growth and survival of *Clostridium perfringens* type A isolates carrying chromosomal or plasmid-borne enterotoxin genes. *Applied and Environmental Microbiology,* 2006, (December), 72 (12).

Juneja V.K., Thippareddi H., Friedman M. Control of *Clostridium perfringens* in cooked ground beef by carvacrol, cinnamaldehyde, thymol, or oregano oil during chilling. *Journal of Food Protection,* 2006, (July), 69 (7).

Novak J.S., Peck M.W., Juneja V.K., Johnson E.A. *Clostridium botulinum* and *Clostridium perfringens*, in *Foodborne pathogens: microbiology and molecular biology.* Eds Fratamico P.M., Bhunia A.K., Smith J.L. Caister. Wymondham,Academic Press. 2005, 383-407.

Montville T.J., Matthews K.R. *Clostridium perfringens,* in *Food microbiology: an introduction.* Eds Montville T.J., Matthews K.R. Washington D.C, ASM Press. 2005, 201-211.

Food Standards Agency, Peck M.W. Improved control of *Clostridium perfringens*. London. FSA, 2004.

de Jong A.E.I., Rombouts F.M., Beumer R.R. Behavior of *Clostridium perfringens* at low temperatures. *International Journal of Food Microbiology,* 2004, (December 1), 97 (1), 71-80.

Wrigley D.M. *Clostridium perfringens*. Foodborne disease handbook, volume 1: bacterial pathogens. Eds Hui Y.H., Pierson M.D., Gorham J.R. 2nd edition New York. Marcel Dekker, 2000, 139-68.

Labbe R.G. *Clostridium perfringens*. The microbiological safety and quality of food, volume 2. Eds Lund B.M., Baird-Parker T.C., Gould G.W. Gaithersburg. Aspen Publishers, 2000, 1110-35.

Bates J.R. *Clostridium perfringens,* in *Foodborne Microorganisms of Public Health Significance.* Ed. Australian Institute of Food Science and Technology. 5th edition North Sydney. AIFST, 1997, 407-28.

Methods of detection

dela Cruz W.P., Gozum M.M.A., Lineberry S.F., Stassen S.D., Daughtry M., Stassen N.A., Jones M.S., Johnson O.L. Rapid detection of enterotoxigenic *Clostridium perfringens* by real-time fluorescence resonance energy transfer PCR. *Journal of Food Protection,* 2006, (June) 69(6).

Wise M.G., Siragusa G.R. Quantitative detection of *Clostridium perfringens* in the broiler fowl gastrointestinal tract by real-time PCR. *Applied and Environmental Microbiology,* 2005, (July) 71 (7).

Labbe R.G. *Clostridium perfringens*. Compendium of methods for the microbiological examination of foods. Eds American Public Health Association, Downes F.P., Ito K. 4th edition. Washington DC. APHA, 2001, 325-30.

Eisgruber H., Schalch B., Sperner B., Stolle A. Comparison of four routine methods for the confirmation of *Clostridium perfringens* in food. *International Journal of Food Microbiology,* 2000 (June 10), 57 (1-2), 135-40.

Public Health Laboratory Service. Enumeration of *Clostridium perfringens,* in *PHLS Standard Methods for Food Products.* Public Health Laboratory Service, London. PHLS, 1999 5.

Schalch B., Eisgruber H., Geppert P., Stolle A. Comparison of four routine procedures for the confirmation of *Clostridium perfringens* from food. *Archiv für Lebensmittelhygiene,* 1996, 47 (1), 27-30.

Rhodehamel E.J., Harmon S.M. *Clostridium perfringens,* in *Bacteriological Analytical Manual.* Ed. Food and Drug Administration. 8th edition. Gaithersburg. AOAC International, 1995, 6.

ENTEROBACTER SAKAZAKII

The organism, until 1980, was classified as the 'yellow-pigmented' *Enterobacter cloacae*. It has since been renamed *Enterobacter sakazakii*, after the Japanese bacteriologist Riichi Sakazaki .The change in classification is based on improved identification methods. *E. sakazakii* is considered an opportunistic pathogen and some strains are known to produce an entero-toxin/endo-toxin (1,2).

The Organism *Enterobacter sakazakii*

Enterobacter sakazakii is a Gram-negative, motile, non-sporing rod, with a facultative anaerobic metabolism, belonging to the family Enterobacteriaceae. It has a biochemical profile similar to *E. cloacae*, but is usually D-sorbitol negative. Other distinguishing characteristics include greater pigment production at temperatures less than 36 °C and utilisation of citrate as the sole source of carbon.

Two morphologically different colony types have been observed, broadly as 'matt' or 'glossy'.

Strains currently classified as *E. sakazakii* fall into two distinct groups; these are further subdivided into 57 strains based on DNA hybridisation, antibiotic susceptibility and biochemical reactions. (3,4,5).

E. sakazakii food poisoning

Despite its ubiquity in nature, *E. sakazakii* has not equalled *Salmonella's* reputation as a pathogen, because the number of outbreaks reported is relatively low. *E. sakazakii* is a frequent cause of nosocomial disease. It has been implicated in sporadic outbreaks of neonatal diseases due to the consumption of contaminated infant formula. Although fresh produce has not been implicated in *E. sakazakii* food poisoning, its ability to survive and grow on fresh produce raises some concern (6).

Incubation time

Challenge studies involving intra-peritoneal injections of suckling mice at a dose of 10^8 cfu/mouse showed that death occurred within 3 days after dosing, typically 24 to 48 h (7).

According to published studies infections are reported within the first day, and up to 18-30 days after birth in premature babies (8).

Symptoms

E. sakazakii induced neonatal meningitis results in ventriculitis, brain abscess or cyst formation and the development of hydrocephalus. Another clinical manifestation is the development of neonatal necrotising enterocolitis (NEC) characterised by intestinal necrosis and pneumatosis intestinalis.

There are only a few reports of the infection among adults caused due to serious underlying diseases such as malignancies. *E. sakazakii* has been obtained from patients suffering from bacteraemia and osteomyelitis, but no reported cases of meningitis (3,4,5).

Mortality

Neonatal meningitis and necrotising enterocolitis is the most common cause of mortality in neonates. Mortality rates are between 40%-80% (1,8).

Infective dose

Since no quantitative determinations of the ingested *E. sakazakii* have been performed the dose-response curve is not known. The effect is thought to be linear to the dose.

Very low initial counts of 1 cfu/ml in bottled infant formula have led to counts of 10^7 per serving of 100 ml in bottles stored at room temperature for 10 h, even sooner in formula held at 35-37 °C (3,5).

Infection

Species of *Enterobacter* are considered opportunistic pathogens and rarely cause disease in otherwise healthy individuals. The bacterium has been implicated most frequently in causing illness in neonates and children from 3 days to 4 years of age. At least 76 cases of *E. sakazakii* infection and 19 deaths in infants have been reported. Nine cases of *E. sakazakii* infection in adults have been reported (4).

E. sakazakii has been documented to cause meningitis, sepsis and necrotizing enterocolitis (9). Pathogenesis in neonates frequently involves bacteraemia and/or sepsis, cerebrospinal fluid (CSF) infection and meningitis with other neurological complications, and necrotizing enterocolitis.

Incidence of *E. sakazakii* in Food Poisoning

The number of documented cases and outbreaks of *E. sakazakii* infection are few. The first cases of *E. sakazakii* were in 1958 in St Albans, England where two cases of terminal neonatal meningitis were reported. The next distinctive report of *E. sakazakii*-induced meningitis infection was in Denmark. The child survived meningitis but experienced severe mental and neurological impairment (4). Additional cases have since been reported and described in America, Netherlands, Greece, Iceland, Belgium, Israel and Germany where *E. sakazakii* was responsible for non-meningital bacteraemia, necrotizing enterocolitis, neonatal septicaemia and *E. sakazakii*-induced meningitis (4,10). In some cases, even though the child survived infection, they were severely mentally retarded or suffered from seizure disorders.

Sources

E. sakazakii sources have been investigated for only a few cases of reported illness, thus the exact reservoir for the organism is unknown. The organism is considered to be ubiquitous.

Clinical

E. sakazakii has been found in the human and animal gut (9). Most isolates are rare and have originated from clinical sources like blood, CSF, sputum, lower and upper respiratory tract, digestive tract, superficial wounds and urine (3,4,5).

Environment

Little is known about the presence of the organism in the environment but it is suspected that water, soil and vegetables are the primary sources of contamination, with flies and rodents secondary means of contamination (4,11).

E. sakazakii has also been isolated from production environments of food factories (milk powder, chocolate, cereal, potato and pasta), and households (3).

Foods

The foods most commonly associated with *E. sakazakii* contamination are dried milk, dried infant foods and powdered infant formula milk. However, *E. sakazakii* has been isolated from a wide range of foods including cheese products, khamir (a fermented bread), sobia (a fermented beverage), minced beef, sausage meat, poultry, grain, various dry food ingredients (e.g. herbs and spices), raw lettuce and other vegetables, (3,9) and drinking water (4).

Growth Survival Characteristic of the Organism in Foods

Temperature

The minimum growth temperature is between 5.5 °C and 8 °C. The lowest recorded temperature that allowed growth of *E. sakazakii* was 3.4 °C, suggesting that the organism is able to grow during refrigeration. The maximum growth temperature ranges from 41-45 °C, in general.

However, under laboratory conditions, 22 strains examined were all capable of growing in brain heart infusion broth at 47 °C (1,3,4,9,12).

pH

Like other members of the Enterobacteriaceae, *E. sakazakii* is presumed to have good resistance to low pH. Survival of the organism in acid environments depends on a number of factors such as pH, acidulant identity, acidulant concentrations, temperature, water activity, atmosphere and the presence of other inhibitory compounds (6).

Survival characteristics of 12 strains of *E. sakazakii* in tryptic soy broth, adjusted to pH 3.0 and 3.5 with HCl, revealed that 10 of the 12 strains showed less than a 1 log decline over a 5 h period at 37 °C (4).

Heat resistance

E. sakazakii is considered to be one of the most thermo-tolerant among the Enterobacteriaceae. As *E. sakazakii* can survive at elevated temperatures (45 °C), and has the ability to grow at temperatures up to 47 °C in warm and dry environments such as in the vicinity of drying equipment in factories, it has a competitive advantage when compared to other members of the Enterobacteriaceae. This also favours its growth even when present at low levels. However, *E. sakazakii* does not survive a standard pasteurisation (>60 °C) process (2,11,12,13).

In a study of 10 strains tested, 5 clinical and 5 food isolates, in reconstituted dried infant formula, a mean $D_{60\,°C}$ of 2.5 minutes and z of 5.82 °C was found. The mean $D_{60\,°C}$ for the 5 clinical isolates was 2.15 min and 3.06 min for the 5 food isolates. In another study the $D_{58\,°C}$ of 12 strains in rehydrated infant formula ranged from 30.5-591.9 s (0.51-9.87 min). When the most heat-resistant strain (z=5.6 °C) was added to a dry infant formula rehydrated at 70 °C, greater than a 4 log reduction was achieved. If 1 cfu/100g of *E. sakazakii* is present in infant formula, a 4-D treatment should ensure absence of the bacterium after cooling (1,12).

Water activity/salt

E. sakazakii can survive in dried infant formula having water activity of *ca* 0.2. Like many other organisms, *E. sakazakii* is relatively resistant in the stationery phase to osmotic and dry stress.

The bacteria protect themselves to increasing osmolarity by the rapid intracellular accumulation of ions, mainly K+ (12).

Control of *E. sakazakii* in Foods

The occurrence of *E. sakazakii* can be reduced through a combination of control measures, for example, the implementation of more stringent hygiene during manufacture, including prevention of recontamination after processing through the environment. Other measures include filtering the air that enters the processing area, keeping overpressure in rooms around the dryer and minimising the use of water for cleaning.

Adherence to hygienic rules during preparation and handling up to consumption is equally important. When preparing and handling infant formula, care should be taken to prevent recontamination through the environment or by improperly sanitised utensils. In addition, holding time (i.e. the time between preparation and consumption) and 'hang time'(i.e. the amount of time the formula is held at room temperature or in the bottle warmer) should be minimised (5,8,14).

In hospitals and maternity units, where central milk kitchens supply prepared bottled feeds for distribution, milk should be sterilised in bottles with the teat already in place. Sterile water should be used for the distribution of feeds (15).

Bibliography

References

1. Jay J.M., Loessner M.J., Golden D.A. Viruses and Some Other Proven and Suspected Foodborne Biohazards, in *Modern Food Microbiology - Seventh Edition.* USA. Springer, 2005, 732.

2. Grant I.R., Houf K., Cordier J-L., Stephan R., Becker B., Baumgartner A. *Enterobacter sakazakii. Mitteilungen aus Lebensmitteluntersuchung und Hygiene*, 2006, 97(1), 22-7.

3. Lehner A., Stephan R. Microbiology, epidemiology and food safety aspects of *Enterobacter sakazakii. Journal of Food Protection,* 2004, 67(12), 2850-57.

4. Gurler J.B., Kornacki J.L., Beuchat L.R. *Enterobacter sakazakii*: A coliform of increased concern to infant health. *International Journal of Food Microbiology,* 2005, 104(1), 1-34.

5. Nazarowec-White M., Faber J.M., Reij M.W., Cordier J.L., van Schothorst M. *Enterobacter sakazakii,* in *Foodborne Pathogens.* Eds Miliotis M.D., Bier J.W. New York. Marcel Dekker, Inc., 2003, 407-14.

6. Kim H., Ryu J.-H., Beuchat L.R. Survival of *Enterobacter sakazakii* on fresh produce as affected by temperature, and effectiveness of sanitizers for its elimination. *International Journal of Food Microbiology,* 2006, (September 1), 111 (2), 134-43.

7. Pagotto F.J., Nazarowec-White M., Bidawid S., Farber J.M. *Enterobacter sakazakii*: infectivity and enterotoxin production in vitro and in vivo. *Journal of Food Protection,* 2003, 66(3) 370-5.

8. Motarjemi Y. Chronic sequelae of foodborne infections, in *Foodborne Pathogens-Hazards, risk analysis and control.* Eds Blackburn C. de W., McClure P.J. England. Woodhead Publishing Ltd., 2002, 505.

9. Anon. Behaviour of *E. sakazakii* on Produce and in Juice. *At a Glance,* 2005, 14(3), 1.

10. Joppen L. *Enterobacter sakazakii*: less well known pathogen takes centre stage. *Food Engineering and Ingredients,* 2005, 30(3), 14-16.

11. Baxter P. Have you heard of *Enterobacter sakazakii*? *Journal of the Association of Food and Drugs Officals*, 2005, 69(1), 16-17.

12. Breeuwer P., Lardeau A., Peterz M., Joosten H.M. Desiccation and tolerance of *Enterobacter sakazakii. Journal of Applied Microbiology,* 2003, 95(3), 967-73.

13. Deseo J. Emerging pathogen: *Enterobacter sakazakii. Inside Laboratory Management*, 2003, 7(3), 32-4.

14. Farber J. *Enterobacter sakazakii* – new foods for thought? *Lancet,* 2004, 363 (9402), 5-6.

15. Roberts D., Greenwood M. Practical Food Microbiology-Third Edition. Massachusetts. Blackwell Publishing Ltd., 2003.

Further reading

Townsend S., Barron J.C., Loc-Carrillo C., Forsythe S. The presence of endotoxin in powdered infant formula milk and the influence of endotoxin and *Enterobacter sakazakii* on bacterial translocation in the infant rat. *Food Microbiology,* 2007, (February), 24 (1), 67-74.

Kim H., Ryu J.-H., Beuchat L.R. Attachment of and biofilm formation by *Enterobacter sakazakii* on stainless steel and enteral feeding tubes. *Applied and Environmental Microbiology,* 2006, (September), 72 (9), 5846-56.

Estuningsih S., Kress C., Hassan A.A., Akineden O., Schneider E., Usleber *E. Enterobacteriaceae* in dehydrated powdered infant formula manufactured in Indonesia and Malaysia. *Journal of Food Protection,* 2006, (December), 69 (12), 3013-17.

Mramba F., Broce A., Zurek L. Isolation of *Enterobacter sakazakii* from stable flies, Stomoxys calcitrans L. (Diptera: Muscidae). *Journal of Food Protection,* 2006, (March), 69 (3), 671-3.

Chaves-Lopez C., De Angelis M., Martuscelli M., Serio A., Papaella A., Suzzi G. Characterization of the *Enterobacteriaceae* isolated from an artisanal Italian ewe's cheese (Pecorino Abruzzese). *Journal of Applied Microbiology,* 2006, (August), 101 (2), 353-60.

Drudy D., O'Rourke M., Murphy M., Mullane N.R., O'Mahony R., Kelly L., Fischer M., Sanjaq S., Shannon P., Wall P., O'Mahony M., Whyte P., Fanning S. Characterization of a collection of *Enterobacter sakazakii* isolates from environmental and food sources. *International Journal of Food Microbiology,* 2006, (July 15), 110 (2), 127-34.

Lee J.W., Oh S.H., Kim J.H., Yook H.S., Byun M.W. Gamma radiation sensitivity of *Enterobacter sakazakii* in dehydrated powdered infant formula. *Journal of Food Protection,* 2006, (June), 69 (6), 1434-7.

Lehner A., Riedel K., Rattei T., Ruepp A., Frishman D., Breeuwer P., Diep B., Eberl L., Stephan R. Molecular characterization of the alpha-glucosidase activity in *Enterobacter sakazakii* reveals the presence of a putative gene cluster for palatinose metabolism. *Systematic and Applied Microbiology,* 2006, (December), 29 (8), 609-25.

Edelson-Mammel S., Porteous M.K., Buchanan R.L. Acid resistance of twelve strains of *Enterobacter sakazakii,* and the impact of habituating the cells to an acidic environment. *Journal of Food Science,* 2006, (August), 71(6), M201-M207.

International Commission on Microbiological specifications for Foods. *Micro-organisms in Foods-Sixth Edition-Microbial Ecology of Food Commodities.* London. Kluwer Academic/Plenum Publishers, 2005.

Lehner A., Riedel K., Eberl L., Breeuwer P., Diep B., Stephan R. Biofilm formation, extra-cellular polysaccharide production, and cell-to-cell signalling in various *Enterobacter sakazakii* strains: aspects promoting environmental persistence. *Journal of Food Protection,* 2005, 68 (11), 2287-94.

Lehner A., Stephan R. Microbiology, epidemiology and food safety aspects of *Enterobacter sakazakii. Journal of Food Protection,* 2004, 67(12), 2850-7.

Nazarowec-White M., Farber J. Incidence, Survival, and Growth of *Enterobacter sakazakii* in Infant Formula. *Journal of Food Protection,* 1997, 60(2), 226-30.

http://www.cfsan.fda.gov/~comm/mmesakaz.html

http://www.nzfsa.govt.nz/consumers/food-safety-topics/recalls-and-product-advice/infant-formula-sakazakii/

http://www.who.int/foodsafety/publications/micro/summary.pdf

www.cdc.gov/mmwr/preiew/mmwrhtml/mm5114a1.htm

www.cfsan.fda.gov/~dms/inf-ltr3.html

Methods of detection

Besse N.G., Leclercq A., Maladen V., Tyburski C., Lombard B. Evaluation of the International Organization for Standardization-International Dairy Federation (ISO-IDF) draft standard method for detection of *Enterobacter sakazakii* in powdered infant food formulas. *Journal of AOAC International,* 2006, (September-October), 89 (5), 1309-16.

Restaino L., Frampton E.W., Lionberg W.C., Becker R.J. A chromogenic plating medium for the isolation and identification of *Enterobacter sakazakii* from foods, food ingredients and environmental sources. *Journal of Food Protection,* 2006, (February), 69 (2), 315-22.

Mullane N.R., Murray J., Drudy D., Prentice N., Whyte P., Wall P.G., Parton A., Fanning S. Detection of *Enterobacter sakazakii* in dried infant milk formula by cationic-magnetic-bead capture. *Applied and Environmental Microbiology,* 2006, (September), 72 (9), 6325-30.

LISTERIA MONOCYTOGENES

Until 1927, the organism was called *Bacterium monocytogenes* and then renamed *Listeria monocytogenes*.

In England and Wales, listeriosis is a comparatively rare disease, and the reported incidence is now about 2-3 cases per million of the population, following a peak in 1988. In recent years, an average total of approximately 100 cases has been reported annually (1), although the figure for 2001 was 136, the highest total since 1989 (source: CDSC; Food Hygiene Laboratory). In the USA, preliminary data from the Foodborne Diseases Active Surveillance Network (FoodNet) indicates an overall incidence of 3 cases per million of the population for 2001. However, because of the potential severity of the disease, food microbiologists need to be well aware of the characteristics of the causative organism and measures for its control in foods. It is generally believed that the majority of cases are foodborne and may be preventable.

The Organism *Listeria monocytogenes*

Listeria are catalase positive, oxidase negative, short Gram-positive, rods with rounded ends, which are non-sporing and have a facultative anaerobic metabolism. They are motile with motility manifesting itself at 20-25 °C; motility is absent at 37 °C. *Listeria monocytogenes* is a species belonging to the genus of bacteria that also includes *L. innocua*, *L. welshimeri*, *L. seeligeri*, *L. ivanovii* and *L. grayi*. *L. innocua* and *L. grayi* are considered to be non-pathogenic, and although *L. ivanovii*, *L. seeligeri* and *L. welshimeri* have very rarely caused human infection (2), almost all cases have been recognised as being due to *L. monocytogenes*.

L. monocytogenes can be divided by serotyping into 13 different serotypes, all of which can cause human listeriosis. However, this is of limited epidemiological value because the majority of cases are caused by serotypes 1/2a, 1/2b, and 4b, and the majority of significant recorded outbreaks have been caused by serotype 4b.

Listeria monocytogenes Food Poisoning

Listeria monocytogenes was discovered almost 100 years ago, but it has only been recognised as an important foodborne pathogen since the early 1980s.

Listeriosis is a generic term for a variety of syndromes caused by *L. monocytogenes*. Many factors have been shown to affect the pathogenicity of *L. monocytogenes*, including the production of a haemolysin and two phospholipases. These virulence mechanisms have been extensively reviewed (2).

Incubation time

The incubation period can vary from 3-70 days (for reasons that are not understood). Symptoms of the gastroenteritic illness occur within hours - usually 18-27 h (3).

Symptoms

Listeriosis most typically takes the form of meningitis, septicaemia and meningoencephalitis, which occurs most commonly in immunosuppressed middle-aged or elderly people. In addition, a flu-like illness can occur in pregnant women, which leads to infection of the foetus and, in turn, can result in miscarriage, stillbirth or birth of a severely ill infant. Currently, about 25% of the cases of listeriosis in England and Wales are pregnancy-associated, but this figure can vary significantly from year to year. Only a small proportion (about 15% in England and Wales) of cases of listeriosis occur amongst people with no known risk factors. Outbreaks have been reported where the principal symptoms were those of gastroenteritis, rather than classical listeriosis (4,5).

Mortality

The overall mortality rate amongst cases of listeriosis is about 30%, but may be as high as 40% in susceptible individuals (6).

Infective dose

The infective dose for listeriosis is uncertain. Food recovered from the homes of a small number of patients who developed listeriosis has been generally heavily contaminated (>10^3 cfu/g), and the typical infective dose is likely to be high (7). However, frankfurters implicated in an outbreak in the US in 1998 were reported to contain less than 0.3 cfu/g (8), and it has been suggested that this may have been an indication of a strain with enhanced virulence (8). The potentially long incubation period for listeriosis causes difficulties in the determination of an infective dose.

The infective dose for the gastroenteritic illness is higher (1.9x 10^5 to 10^9 cfu/g) (3).

Foods involved

Many cases of listeriosis have not been directly related to a known food source, and it is often difficult to differentiate between cases that are food-linked and those that are not, again because of the long incubation period involved.

The first outbreak of human listeriosis where food was convincingly implicated occurred in 1981, involving coleslaw salad in an outbreak in Canada, with at least 41 cases and 7 deaths. In this outbreak, cabbage had become contaminated with *Listeria* from infected sheep manure, which was used as fertiliser in the soil in which the cabbages were grown. The manure used was from sheep suspected of suffering from listeriosis (9).

At least 11 further major outbreaks of human foodborne listeriosis have occurred: a possibly milkborne outbreak in Boston, USA; 'Mexican-style' cheese in Los Angeles, USA; 'Vacherin Mont d'Or' cheese in Switzerland; pâté in the UK; and pork tongue in aspic in France (7). Also 'rillettes' (potted pork) were implicated in a further outbreak involving at least 25 cases in mid-1993 in France (7). Chocolate milk caused an outbreak of 45 cases in the USA in 1994 (5), and a raw milk soft cheese was implicated in a 1995 outbreak in France (7). More recently, contaminated frankfurters were associated with a multi-state outbreak of over 100 cases in the US that took place between August 1998 and March 1999, during which 15 adult deaths and 6 miscarriages were recorded (8). An outbreak in Finland in 1998-99 was linked to butter (10), pork tongue in jelly was implicated in an outbreak in France in 1999-2000 (11), and an outbreak in the US in 2000 was associated with ready-to-eat turkey deli meat (12).

Incidence of *L. monocytogenes* Food Poisoning

The incidence of listeriosis worldwide increased dramatically throughout the early to mid-1980s. In 1988, numbers in England and Wales peaked at nearly 300, possibly as a result of a single foodborne outbreak (13). However, despite rigorous surveillance, numbers have since seen a decline to a level similar to that in the early 1980s. The decline in cases coincided with the issue of UK Government health warnings, concerned with the consumption of pâté.

Similar trends were observed in France and the USA. Before 1974, there were about 15 cases per year in France compared with 687 cases in 1987. In the US, during 1967-69 there were 255 cases compared with about 1,700 cases in 1987 (14). In Canada, between 1989 and 1990, 24% of food poisoning incidents were caused by *L. monocytogenes,* but this declined to ≤3% during 1991-1992 (3).

Sources

Humans

Person-to-person transmission of infection has been documented as person-to-person spread in hospitals, but only during the neonatal period.

Surveys into the carriage rate of *L. monocytogenes* in healthy humans have shown a wide range in percentage carriage rates; 5-10% is a figure that is commonly mentioned.

Animals and the environment

Infection from *Listeria* can originate from direct or indirect contact with animals (sheep, goats, and cows can excrete *L. monocytogenes*, in faeces or sometimes in milk). *L. monocytogenes* is also ubiquitous in the environment, and survives and grows in soil and water, so it can be transferred to foods from a wide range of sources. However, evidence for tracing the transmission of human listeriosis suggests that post-process contamination within the food manufacturing unit, usually from plant or machinery, is of particular importance (15).

Pasture grass and silage - especially silage of poor quality - are important sources of *Listeria* in ruminant livestock.

Foods

Foods can become contaminated with *Listeria* at any stage in the food chain, from the farm, through processing and distribution, to the consumer's kitchen, especially in moist/wet environments. Numerous studies have now indicated that *L. monocytogenes* can be found in a very wide range of foods, including milk, soft cheese, raw and pre-cooked chicken and meats, pâté, fermented sausage, vegetables, smoked and lightly processed fish products (16), and seafoods (2,3,17). In the UK, *L. monocytogenes* has been found in many foods, usually in very low numbers, but occasionally in numbers exceeding 10,000/g, particularly in pâté and soft cheese (18). Most foods where *L. monocytogenes* has been found have contained serotype 1/2a, b or c. However, pâté has frequently been found to contain serotype 4b - the serotype most commonly isolated from humans (18). Cooked, chilled, ready-to-eat meat, poultry, and sausage products are now of particular concern, and *L. monocytogenes* has been isolated from samples of these foods in many countries (16).

Although *Listeria* spp other than *L. monocytogenes* are not a cause of human listeriosis, their presence in foods, particularly those that have undergone a listericidal process, may be indicative of poor hygiene practices.

There is evidence that the microbiological quality of some foods, with respect to contamination by *Listeria,* improved during the late 1980s, possibly as a result of changes in UK legislation and implementation of better hygiene standards by the food industry (13,19).

Foods associated with transmission of infection are diverse; processed meat, dairy, vegetable or marine products have all been implicated. However, these generally have been able to support the growth of *L. monocytogenes*, and have a high degree of processing, and an extended shelf-life, often at chill temperatures.

Growth/Survival Characteristics of the Organism in Foods

Temperature

Listeria monocytogenes is unusual amongst foodborne pathogens in that it is psychrotrophic, being potentially capable of growing at refrigeration temperatures down to, or even below, 0 °C. Growth at low temperatures can be very slow, requiring days to double in number. As part of a mixed inoculum with other psychrotrophic organisms, *L. monocytogenes* was found to grow in vacuum-packed sliced roast beef at a temperature of -1.5 °C, but with a lag time of 174 days and a generation time of 100 h (20). However, -0.4 °C is probably a more likely minimum in foods (21). Its optimum growth temperature is between 30 °C and 37 °C.

The upper temperature limit for the growth of *L. monocytogenes* is reported to be 45 °C. *L. monocytogenes* is killed at temperatures of >50 °C (3).

The organism survives well for several weeks in frozen foods (22), but less well under acid conditions.

Heat resistance

Listeria monocytogenes is not a particularly heat-resistant organism; it is not a spore-former, so can be destroyed by pasteurisation. It has, however, been reported to have slightly greater heat resistance than certain other foodborne pathogens such as *Salmonella*. It is generally agreed that commercial milk pasteurisation will destroy normal levels of *L. monocytogenes* in milk (>10^5/ml). In milk at 62.8 °C for 30 min 39D and at 71.7 °C for 15 sec 5.2D, reductions have been obtained. The UK Department of Health advised that ready meals or similar products should receive a heat treatment of at least 2 min at 70 °C or equivalent to ensure the destruction of *Listeria*. In the US, standards for the heat treatment of frozen dessert mixes require a process of 68.3 °C for 30 min, or 79.4 °C for 25 s to ensure destruction of *L. monocytogenes*. The addition of stabilisers such as guar gum and carrageenan to ice cream mixes is thought to increase the heat resistance of the organism in these products (23).

Typical D-values in foods range between approximately 5-8 min at 60 °C and 0.1-0.3 min at 70 °C, depending on strain and substrate. Heat shocked cultures have been reported to show enhanced heat resistance (24).

pH

The ability of *Listeria* to grow at different pH values (like other bacteria) is markedly affected by the type of acid used and temperature. Under otherwise ideal conditions, the organism is able to grow at pH values well below pH 5; pH 4.3 is the lowest value where growth has been recorded, using hydrochloric acid as acidulant. In foods, however, the lowest limit for growth is likely to be considerably higher - especially at refrigeration temperatures, and where acetic acid is used as acidulant; pH <5.2 has been indicated as inhibitory for growth in foods (16). Nevertheless, *L. monocytogenes* is relatively acid tolerant, and has been shown to survive for 21 days in orange serum adjusted to pH 3.6 and incubated at 4 °C (25).

Water activity/salt

Listeria monocytogenes is quite tolerant of high sodium chloride/low water activities. It is likely to survive, or even grow, at salt levels found in foods of 10-12% NaCl or more. It grows best at a_w of ≥0.97, but has been shown to be able to grow at a_w level of 0.90. The bacterium may survive for long periods at a_w values as low as 0.83. (3). Tolerance of water activity levels is influenced by temperature, pH and type of solute (a slightly lower a_w minimum - 0.90 has been observed with glycerol rather than NaCl or sucrose at 30 °C) (16).

Atmosphere

L. monocytogenes grows well under aerobic and anaerobic conditions, and a study using broth culture showed that anaerobic incubation favoured growth (26). Also, several studies with modified atmospheres showed that *L. monocytogenes* grew at low temperatures (3-7 °C) on meat in reduced-oxygen or anaerobic conditions (27,28). Similarly, *L. monocytogenes* grew at low temperatures (5 °C) on various vegetables stored under modified atmospheres (3-10% CO_2) (29). High concentrations (80%) of CO_2 have been found to be necessary to inhibit growth (30). Studies have also shown that *L. monocytogenes* is not significantly affected by vacuum-packing (3).

Irradiation

L. monocytogenes cells are more resistant to radiation than are the cells of Gram-negative pathogens such as *Salmonella*. Studies of the resistance of *L. monocytogenes* to gamma irradiation in selected products gave a range of D-values of 0.38–1.00 kGy (16). Resistance is generally greater in foods than in culture medium (31). Treatments of 1.7-4.0 kGy have been found sufficient to give a 6-7 log reduction in numbers of *Listeria* spp (32).

It has been reported that irradiation treatments were significantly more lethal to *L. monocytogenes* in turkey meat packed in the presence of air than in MAP or vacuum-packed meat, suggesting increased sensitivity in the presence of oxygen (32).

Summary of Control of *L. monocytogenes* in Foods

The WHO's Informal Working Group on Foodborne Listeriosis in 1988 concluded that "the total elimination of the organism from all food is impractical and may be impossible" (33). This is still a generally accepted view. The common occurrence of *L. monocytogenes* in the environment and in foods indicates that exposure to *L. monocytogenes* does not usually lead to infection. The incidence of listeriosis, in comparison with other foodborne illness, such as salmonellosis, is very rare. Nevertheless, the potential severity of listeriosis demands that every effort should be made to reduce the incidence of the organism in foods to a minimum. In addition, the presence of *Listeria* in a food - especially a cooked food - may be an indication of poor hygiene during manufacture, distribution or retailing. The United States Department of Agriculture (USDA) has adopted a 'zero tolerance policy' for *L. monocytogenes* in ready-to-eat foods, but in Europe counts of up to 100 CFU/g at end of shelf-life are generally considered acceptable.

The HACCP approach to food quality assurance should be applied at all levels in the food industry, with particular attention being paid to temperature control, and the strict adherence to practices that prevent contamination, especially cross-contamination between raw and processed foods, and wet processing/manufacturing or packaging environments and the processed foods. Storage times for sensitive foods should also be carefully controlled.

The UK Department of Health advises vulnerable individuals, especially pregnant women and the immunosuppressed, to take appropriate precautions against *Listeria* infection, including avoiding eating pâté and soft (mould-ripened) cheeses, and to reheat cook-chill foods adequately (18). In the US, recent advice from the FDA adds chilled ready-to-eat meats, hot dogs, and smoked seafood, unless properly reheated, to the list of foods that 'at risk' consumers should avoid.

Bibliography

References

1. Anon. Listeriosis in England and Wales: 1983 to 1996. *CDR Weekly,* 1997, 7 (11), 95.

2. Pagotto F., Corneau N., Farber J. *Listeria monocytogenes* infections, in *Foodborne Infections and Intoxications.* Eds Riemann H.P., Cliver D.O. London. Elsevier, 2006, 313-31.

3. Montville T.J., Matthews K.R. *Listeria monocytogens,* in *Food Microbiology - An Introduction.* Washington. ASM Press, 2005, 129-38.

4. Salamina G., Dalle Donne E., Niccolini A., Poda G., Cesaroni D., Bucci M., Fini R., Maldini M., Schuchat A., Swaminathan B., Bibb W., Rocourt J., Binkin N., Salmaso S. A foodborne outbreak of gastroenteritis involving *Listeria monocytogenes. Epidemiology and Infection,* 1996, 117 (3), 429-36.

5. Dalton C.B., Austin C.C., Sobel J., Hayes P.S., Bibb W.F., Graves L.M., Swaminathan B., Proctor M.E., Griffin P.M. An outbreak of gastroenteritis and fever due to *Listeria monocytogenes* in milk. *New England Journal of Medicine,* 1997, 336 (2), 100-5.

6. Rocourt J. Risk factors for listeriosis. *Food Control,* 1996 7 (4-5), 195-202.

7. McLauchlin J. The relationship between *Listeria* and listeriosis. *Food Control,* 1996, 7 (4-5), 187-93.

8. Mead P.S. Multistate outbreak of listeriosis traced to processed meats, August 1998-March 1999. *Epidemiological Investigation Report,* 1999, Centers for Disease Control and Prevention, Atlanta.

9. Schlech W.F.III, Lavigne P.M., Bortolussi R.A., Allen A.C., Haldane E.V., Wort A.J., Hightower A.W., Johnson S.E., King S.H., Nicholls E.S., Broome C.V. Epidemic listeriosis - evidence for transmission by food. *New England Journal of Medicine,* 1983, 308, 203-6.

10. Lyytikainen O., Autio T., Maijala R., Ruutu P., Honkanen-Buzalski T., Miettinen M., Hatakka M., Mikkola J., Anttila V.-J., Johansson T., Rantala L., Aalto T., Korkeala H., Siitonen A. An outbreak of *Listeria monocytogenes* serotype 3a infections from butter in Finland. *Journal of Infectious Diseases*, 2000, 181, 1838-41.

11. Anon. Outbreak of *Listeria monocytogenes* serotype 4b infection in France. *CDR Weekly*, 2000, 10 (81), 84.

12. Hurd S. *et al.* Multistate outbreak of Listeriosis - United States, 2000. *Morbidity and Mortality Weekly Report (MMWR),* 2000, 49 (50), 1129-30.

13. McLauchlin J., Hall S.M., Velani S.K., Gilbert R.J. Human listeriosis and pâté: a possible association. *British Medical Journal,* 1991, 303 (6805), 773-5.

14. Klima R.A., Montville T.J. The regulatory and industrial responses to listeriosis in the USA: a paradigm for dealing with emerging foodborne pathogens. *Trends in Food Science and Technology,* 1995, 6 (3), 87-93.

15. Rocourt J. *Listeria monocytogenes*: the state of the science. *Dairy, Food and Environmental Sanitation,* 1994, 14 (2), 70, 72-82.

16. Ryser E.T., Marth E.H. *Listeria, listeriosis and food safety*. New York. Marcel Dekker, 2007.

17. Jay J.M. Prevalence of *Listeria* spp. in meat and poultry products. *Food Control,* 1996, 7 (4-5), 209-14.

18. Roberts D. *Listeria monocytogenes* and food: the UK approach. *Dairy, Food and Environmental Sanitation,* 1994, 14 (4), 198-204.

19. Gilbert R.J. Zero tolerance for *Listeria monocytogenes* in foods - is it necessary or realistic? *Food Australia,* 1996, 48 (4), 169-70.

20. Hudson A.J., Mott S.J., Penney N. Growth of *Listeria monocytogenes, Aeromonas hydrophila*, and *Yersinia enterocolitica* on vacuum and saturated carbon dioxide controlled atmosphere-packaged sliced roast beef. *Journal of Food Protection,* 1994, 57 (3), 204-8.

21. Walker S.J., Archer P., Banks J.G. Growth of *Listeria monocytogenes* at refrigeration temperatures. *Journal of Applied Bacteriology,* 1990, 68 (2), 157-62.

22. Kaya M., Schmidt U. Behaviour of *Listeria monocytogenes* in minced meat during chilled and frozen storage. *Fleischwirtschaft,* 1989, 69 (4), 617-20.

23. Farber J.M., Peterkin P.I. *Listeria monocytogenes*, in *The microbiological safety and quality of food, volume 2.* Eds Lund B.M., Baird-Parker T.C., Gould G.W. Gaithersburg. Aspen Publishers, 2000, 1178-1232.

24. International Commission on Microbiological Specifications for Foods. Microorganisms in Foods, Volume 5: Microbiological Specifications of Food Pathogens. London. Blackie, 1996.

25. Parish M.E., Higgins D.P. Survival of *Listeria monocytogenes* in low pH model broth systems. *Journal of Food Protection,* 1989, 52 (3), 144-7.

26. Buchanan R.L., Stahl H.G., Whiting R.C. Effects and interactions of temperature, pH, atmosphere, sodium chloride and sodium nitrite on the growth of *Listeria monocytogenes. Journal of Food Protection,* 1989, 42 (12), 844-51.

27. Garcia de Fernando G.D., Nychas G.J.E., Peck M.W., Ordonez J.A. Growth/survival of psychrotrophic pathogens on meat packaged under modified atmospheres. *International Journal of Food Microbiology,* 1995, 28 (2), 221-31.

28. Beumer R.R., te Giffel M.C., de Boer E., Rombouts F.M. Growth of *Listeria monocytogenes* on sliced cooked meat products. *Food Microbiology,* 1996, 13 (4), 333-40.

29. Berrange M.E., Brackett R.E., Beuchat L.R. Growth of *Listeria monocytogenes* on fresh vegetables stored under controlled atmosphere. *Journal of Food Protection,* 1989, 52 (10), 702-5.

30. Hendricks M.T., Hotchkiss J.H. Effect of carbon dioxide on the growth of *Pseudomonas fluorescens* and *Listeria monocytogenes* in aerobic atmospheres. *Journal of Food Protection,* 1997 (December), 60 (12), 1548-52.

31. Gursel B., Gurakan G.C. Effects of gamma irradiation on the survival of *Listeria monocytogenes* and on its growth at refrigeration temperature in poultry and red meat. *Poultry Science,* 1997 (December), 76 (12), 1661-4.

32. Thayer D.W., Boyd G. Irradiation and modified atmosphere packaging for the control of *Listeria monocytogenes* on turkey meat. *Journal of Food Protection*, 1999, (October) 62 (10), 1136-42.

33. World Health Organization. Foodborne listeriosis: report of a WHO informal working group, Geneva, February 1988. Geneva. WHO, 1988.

Further reading

Oussalah M., Caillet S., Saucier L., Lacroix M. Inhibitory effects of selected plant essential oils on the growth of four pathogenic bacteria: *E. coli* O157:H7, *Salmonella typhimurium*, *Staphylococcus aureus* and *Listeria monocytogenes*. *Food Control,* 2007, (May), 18 (5), 414-20.

Ryser E.T, Marth E.H. *Listeria, listeriosis and food safety*. New York. Marcel Dekker, 2007.

McLauchlin J. *Listeria,* in *Emerging Foodborne Pathogens.* Eds Motarjemi Y., Adams M. Cambridge. Woodhead Publishing Limited, 2006, 406-26.

Swaminathan B. *Listeria monocytogenes*, in *Food Microbiology: Fundamentals and Frontiers.* Eds Doyle M.P., Beuchat L.R., Montville T.J. 2nd edition. Washington DC. ASM Press, 2001, 383-409.

Donnelly C.W. *Listeria monocytogenes*, in *Guide to Foodborne Pathogens.* Labbe R.G., Garcia S. New York. Wiley, 2001, 99-132.

Farber J.M., Peterkin P.I. *Listeria monocytogenes*, in *The Microbiological Safety and Quality of Food, Volume 2.* Eds Lund B.M., Baird-Parker T.C., Gould G.W. Gaithersburg. Aspen Publishers, 2000, 1178-1232.

Kyriakides A., Bell C. *Listeria.* London. Blackie, 1997.

Sutherland P.S., Porritt R.J. *Listeria monocytogenes*, in *Foodborne Microorganisms of Public Health Significance.* Ed. Australian Institute of Food Science and Technology. 5th Edition. North Sydney. AIFST, 1997, 333-78.

Schott W., Hildebrandt G. Overview on *Listeria* in processed meat products, in *Factors Affecting the Microbial Quality of Meat, Volume 3: Cutting and Further Processing.* Eds Commission of the European Communities, Hinton M.H., Rowlings C. Bristol: University of Bristol Press, 1996, 177-94.

International Commission on Microbiological Specifications for Foods. *Listeria monocytogenes*, in *Microorganisms in Foods, Volume 5: Microbiological Specifications of Food Pathogens.* Ed. International Commission on Microbiological Specifications for Foods. London. Blackie, 1996, 141-82.

Farber J.M., Coates F., Daley E. Minimum water activity requirements for the growth of *Listeria monocytogenes. Letters in Applied Microbiology,* 1992, 15 (3), 103-5.

Methods of detection

Ryser E.T., Marth E.H. *Listeria, listeriosis and food safety.* New York. Marcel Dekker, 2007.

Schindler B.D., Shelef L.A. Immobilization and detection of *Listeria monocytogenes. Applied and Environmental Microbiology,* 2006, (June), 72 (6), 4426-8.

Oravcova K., Kaclikova E., Siekel P., Girotti S., Kuchta T. Detection of *Listeria monocytogenes* in food in two days using enrichment and 5-nuclease polymerase chain reaction with end-point fluroimetry. *Journal of Food and Nutrition Research,* 2007, 46 (1), 35-8.

Wiedmann M. Molecular subtyping methods for *Listeria monocytogenes. Journal of AOAC International,* 2002 (March-April), 85 (2), 524-31.

Ryser E.T., Donnelly C.W., Andrews W.H., Flowers R.S., Silliker J., Bailey J.S., Lampel K.A. *Listeria, Salmonella and Shigella*, in *Compendium of Methods for the Microbiological Examination of Foods*. Eds American Public Health Association, Downes F.P., Ito K. 4th edition Washington DC. APHA, 2001, 343-85.

Andrews W.H. Microbiological methods. Subchapter 10. *Listeria* (parts 1 & 2), in Official Methods of Analysis of AOAC International, volume 1: Agricultural Chemicals, Contaminants, Drugs. AOAC International. Ed. Horwitz W. 17th edition. Gaithersburg. AOAC International, 2000, Chapter 17, 138-63.

Batt C.A. Rapid methods for detection of *Listeria*. Listeria, listeriosis and food safety. Eds Ryser E.T., Marth E.H. 2nd edition. New York. Marcel Dekker, 1999, 261-78.

Donnelly C.W. Conventional methods to detect and isolate *Listeria monocytogenes*. Listeria, listeriosis and food safety. Eds Ryser E.T., Marth E.H. 2nd edition. New York. Marcel Dekker, 1999, 225-60.

British Standards Institution. Microbiology of food and animal feeding stuffs. Horizontal method for the detection and enumeration of *Listeria monocytogenes*. Part 2: Enumeration method. BS EN ISO 11290-2:1998. London. BSI, 1998.

MAFF. MAFF validated methods for the analysis of foodstuffs: method for the enumeration of *Listeria monocytogenes* in meat and meat products V38. *Journal of the Association of Public Analysts,* 1997, 33 (2), 67-85.

McLauchlin J. The identification of *Listeria* species. *International Journal of Food Microbiology,* 1997 (August 19), 38 (1), 77-81.

British Standards Institution. Microbiology of food and animal feeding stuffs. Horizontal method for the detection and enumeration of *Listeria monocytogenes*. Part 1. Detection method. BS EN ISO 11290-1:1997. BS 5763: Part 18. London. BSI, 1997.

Arnold G.J., Sutherland P.S., Szabo E.A. *Listeria* methods workshop manual: detection, identification and typing of *Listeria monocytogenes* in foods: proceedings of a workshop, Perth, October 1995 (ISOPOL XII). Uppsala. SLU, 1996.

Hitchins A.D. *Listeria monocytogenes*. Bacteriological Analytical Manual. Ed. Food and Drug Administration. 8th Edition. Gaithersburg: AOAC International, 1995.

Curtis G.D.W., Lee W.H. Culture media and methods for the isolation of *Listeria monocytogenes*. Culture Media for Food Microbiology. Eds Corry J.E.L., Curtis G.D.W., Baird R.M. Amsterdam. Elsevier, 1995, 63-75.

IDF Standard 143A: 1995. Milk and milk products. Detection of *Listeria monocytogenes*. *International Dairy Federation*, 1995.

Dever F.P., Schaffner D.W., Slade P.J. Methods for the detection of foodborne *Listeria monocytogenes* in the US. *Journal of Food Safety,* 1993, 13 (4), 263-92.

Gavalchin J., Landy K., Batt C.A. Rapid methods for the detection of *Listeria*. Molecular Approaches to Improving Food Quality and Safety. Eds Bhatnagar D., Cleveland T.E. New York. Van Nostrand Reinhold, 1992, 189-204.

SALMONELLA

Salmonella is a common cause of food poisoning worldwide. There are many different types of *Salmonella* but, with the exception of the few which cause typhoid or paratyphoid fever, the illness they cause is similar. *Salmonella* are widespread in cows, poultry, eggs, pigs, pets and wild animals.

The Organism *Salmonella*

Salmonellae belong to the family Enterobacteriaceae. They are Gram-negative, non-sporing, oxidase-negative and catalase-positive. Most strains are motile and can grow both aerobically and anaerobically.

The nomenclature of *Salmonella* has been progressively revised on the basis of biochemical and serological characteristics. Recently, a classification based on that proposed by Le Minor & Popoff (1) has been increasingly adopted. On the basis of DNA/DNA hybridisation, it has been proposed that there are only two species of *Salmonella*, *S. enterica*, and *S. bongori*. However, *S. enterica* can be further divided into six sub species. The genus is further divided, serologically, into over 2,400 serovars (or serotypes), these being distinguished according to their possession of different antigens (O, H, or Vi). Most of the recognised serovars belong to the species *S. enterica*, and only 20 are included in *S. bongori*. Nearly 1,500 serovars, including almost all of those important in foodborne disease, belong to a single subspecies, *S. enterica* subsp. *enterica*. The serotype names have traditionally been used as if they denote species (e.g. *S. enteritidis*, *S. typhimurium*), but an example of the correct nomenclature under the currently accepted classification would be *S. enterica* subsp. *enterica* serovar Enteritidis (2). For convenience this can be abbreviated to *S.* Enteritidis, and this form will be continued here. Further differentiation of *Salmonella* isolates can be done by biochemical characteristics and by 'phage-typing'.

There are three different syndromes in man caused by salmonellae. The most severe is enteric (typhoid) fever, caused by the host adapted organisms *Salmonella* Typhi, *S.* Paratyphi A, B (*S.* Schottmuelleri) and C (*S.* Hirschfeldii), and this will not be considered further here. The second syndrome is by far the most frequently encountered, gastroenteritis or food poisoning caused by non-typhoid salmonellae. Finally, a variety of systemic infections (such as septicaemia) and chronic conditions may develop following infection of susceptible individuals by non-typhoid strains. Certain non-human adapted serotypes such as *S.* Dublin, *S.* Enteritidis and *S.* Virchow, can be invasive and give rise to such infections.

The two main serotypes of *Salmonella* that are of current concern in Europe and the United States are *S.* Enteritidis and *S.* Typhimurium. Their importance is due to their frequency of isolation from incidents of food poisoning, and their involvement in the on-going issue of the safety of poultry and eggs (3). However, a large number of other serotypes, particularly *S.* Virchow, have been involved in food poisoning, and all serotypes of *Salmonella* can be considered to be equally significant in relation to food safety. Recently, there has been concern over increasing antibiotic resistance in some salmonellae, such as *S.* Typhimurium DT104. This strain has become a common human pathogen in England and Wales, and has also been isolated from livestock, domestic pets, and wildlife. It has shown an increasing spectrum of antibiotic resistance since 1990 (4), and has recently been documented as developing resistance to fluoroquinolones in England and Wales, and in Germany (5). The World Health Organisation (WHO) reports that Multidrug-resistant strains of *Salmonella* are now frequently encountered, and that this characteristic is an integral part of the genetic material (6).

Salmonella Food Poisoning

Gastroenteritis is the usual manifestation of infection caused by *Salmonella*. After *Campylobacter*, *Salmonella* is the most commonly reported cause of gastroenteritis in the UK, and it continues to be one of the main causes of foodborne illness the world over.

The severity of illness varies according to the strain of *Salmonella* involved and to the susceptibility of the host. The most susceptible are infants, the elderly, and individuals who are immunocompromised.

Individuals recovering from salmonellosis can continue to shed salmonellae in their stools for some time after illness; however, long-term carriage is uncommon.

Incubation time

The incubation period for *Salmonella* food poisoning is between 8 and 72 h.

Symptoms

The symptoms of foodborne *Salmonella* infection include diarrhoea, abdominal pains, chills, fever, nausea and vomiting, as well as dehydration and headache. The symptoms usually last from 2-5 days.

Mortality

Death from salmonellosis is rare (less than 1% of cases), but is more likely amongst the susceptible group outlined above (3).

Infective dose

Until recently, it was considered that the ingestion of large numbers (minimum 10^6) of salmonellae was needed in order for salmonellosis to occur. However, more recent outbreaks have indicated that under certain conditions and in certain foods (e.g. chocolate, or cheese, where high fat content/low water activity appears to protect *Salmonella* from stomach acidity), only small numbers need to be ingested to induce illness in the very young or elderly, possibly as low as 10-100 cells (7,8). The infective dose for a *S.* Enteritidis outbreak, associated with ice cream that occurred in the US in 1994, has been estimated at no more than 25 cells (9).

Foods involved

Types of food involved in foodborne salmonellosis have been wide-ranging, including poultry, meat products, dairy products, eggs, fruit juice, tomatoes, lettuce, canteloupe melons, seed sprouts, toasted oat cereals, paprika flavoured potato crisps, coconut, almonds, peanut butter and chocolate.

Fermented meats such as salami have been implicated in *Salmonella* outbreaks. In the UK, during 1988, there was an outbreak due to *S.* Typhimurium DT124 and 71 cases were confirmed, of which 55 concerned children (10). A very large outbreak of *Salmonella* occurred in the US in 1985; there were over 16,000 cases associated with contaminated pasteurised milk (11). In the same year, in the UK, an outbreak of *S.* Ealing was associated with contaminated milk powder (12). The 1994 *S.* Enteritidis outbreak, associated with ice cream in the US, is estimated to have infected up to 224,000 people, and is probably the largest recorded.

There have been several outbreaks involving chocolate, if chocolate becomes contaminated with *Salmonella* during the production process , or made using contaminated ingredients, the finished product will also be contaminated because the process of chocolate manufacture cannot include a sufficiently severe heat process to kill *Salmonella*.

Incidence of *Salmonella* Food Poisoning

In England and Wales, there was a marked increase in the incidence of foodborne salmonellosis, especially after 1985, reaching a peak of 32,000 reported cases in 1997. This was attributed mainly to the rise in *S.* Enteritidis PT4 infections. In 1993, the UK Advisory Committee on the Microbiological Safety of Food concluded that, between 1981 and 1991, there was a rise of over 170% in reported cases of salmonellosis, primarily because of that increase in *S.* Enteritidis (particularly *S.* Enteritidis PT4). It is probable that this increase was due to its presence in certain foods - in particular, eggs and poultry. However, since 1997, there has been a dramatic decrease in the incidence of infection, and only about 12,500 cases were recorded in 2006 compared to approximately 31,000 in 1997 (source: CDSC). This is thought to be largely attributable to the reduced prevalence of *S.* Enteritidis in eggs and poultry, as a result of action taken by producers.

In November 2003, a large outbreak in the UK (324 cases) was traced to a single source. 19% of the cases were admitted to hospital. *S.* Enteritidis PT56 was found to be causative agent.

A shipment of raw beef was implicated in an outbreak of *Salmonella* Typhimurium DT104 identified in the Netherlands during October-December 2005 (14).

Since 1989, there have been about 30,000 *Salmonella* infections reported each year in England and Wales, and *S.* Enteritidis PT4 accounted for about 40% of these. In 2006, there were under 2,000 reports of *S.* Enteritidis PT4 infections. In 2001, for the first time, *S.* Enteritidis PT4 accounted for less than 50% of all *S.* Enteritidis isolates from humans, as other phage types, such as PT1 and PT6 have become more common (15).

Information worldwide suggests that the incidence of salmonellosis has increased in many countries, and infection caused by *S.* Enteritidis, in particular, has increased in several developed countries.

In European countries, the increase in incidence of *S.* Enteritidis is also due primarily to phage type 4.

There was a considerable increase in salmonellosis in the US in the early 1990s. This was also due largely to *S.* Enteritidis, but in North American countries the predominant phage types are PT8, PT13 and PT13a. However, during the period from 1996-2001, the incidence of salmonellosis decreased by 15% to an incident rate of 15.1 per 100,000 people. During the same period, the incidence of infection by *S.* Enteritidis, and *S.* Typhimurium fell by 22% and 24% respectively, but the incidence of some other serovars, such as *S.* Newport, and *S.* Heidelberg, increased significantly (16).

In the US, it has been estimated that the true annual incidence of food poisoning due to *Salmonella* may be as high as 4.8 million (17). In the UK, a true annual incidence of more than 2 million has been suggested (3), but more recent estimates using the 'Delphi method' produced an annual figure of 537,000 cases (18).

Sources

Humans

Contamination of food by infected food handlers is unusual; it is only likely to occur when a handler has diarrhoea and contaminated hands come in direct contact with food that is not subsequently cooked.

Human transmission via the faecal-oral route can occur, especially among patients in hospitals and institutions. In these cases, poor personal hygiene has a vital role in the transmission.

Animals and environment

Food animals can become infected via direct contact with other infected or symptomless animals, via contaminated feed or water, via their environment, or via wild birds or rodent pests.

Salmonella may also infect common pets such as cats and dogs, and thus these animals may act as a source of the food contamination. Reptiles, such as terrapins, are also common carriers of *Salmonella*.

In addtition, manufacturing, catering, or domestic environments can become contaminated with *Salmonella* and act as a source of the organism, for example, mops and cloths, refrigerators, slicing machines and other inadequately cleaned and sanitised equipment can harbour *Salmonella*.

Foods

The main source of *Salmonella* for man is food from infected food animals. Thus, meat, poultry, eggs, or raw milk can become contaminated with intestinal/faecal material from infected livestock. Eggs can be contaminated on the shell with *S.* Enteritidis, as well as other salmonellas, as a result of either faecal carriage of the chicken or contamination from the environment. Under certain conditions, shell organisms are able to penetrate into egg contents. More importantly, however, egg contents can be contaminated with *S.* Enteritidis as a result of systemic infection of the laying hen, which results in infection of the reproductive tissues, particularly the oviduct. In 1993, it was estimated that in eggs sampled from retail outlets, 1 in 880 were positive for *Salmonella* (19). An unpublished report suggested that eggs sampled in 1995/6 showed a contamination rate of approximately 1 in 700. However, evidence collected by the egg industry more recently suggests that the contamination rate of eggs produced in the UK has fallen significantly since then, as a result of improvements in hygiene, and the large-scale vaccination of laying flocks.

Consumption of contaminated foods that are consumed raw or not properly cooked can lead to food poisoning. Cross-contamination of other food materials that are not subsequently cooked - during processing, via chopping boards and other equipment used during food preparation - can lead to further incidents of food poisoning.

The prevalence of *Salmonella* in foods is variable. In 2001, a UK study reported that an average of 5.8% of poultry carcases were contaminated with *Salmonella* (20). This compares to a figure of about 40% reported in 1996 (21). In Spain, a study of the prevalence in chicken legs carried out in 1999 showed that 35.8% of samples were contaminated (22), and in the same year, a survey of *Salmonella* in broilers in the US gave a contamination rate of 11% (23). The prevalence in beef and lamb carcases is reported to be less than 1% (21). Vegetables, fruit and herbs/spices can be contaminated with

Salmonella, and levels reported are 1.9-8%, ≤5.4% and 6.7-13.8%, respectively (7). Data collected for seafood in the US over an eight-year period during the 1990s showed that the overall prevalence of *Salmonella* was 7.2% for imported products, and 1.3% for domestically produced seafood (24).

Growth/Survival Characteristics of the Organism in Foods

Temperature

Most *Salmonella* serotypes can grow in the temperature range of 7-48 °C. However, some strains are able to grow at temperatures as low as 4 °C (25). Growth is slow at temperatures below about 10 °C, the optimum being 35-37 °C. Although most salmonellae cannot grow at refrigeration temperature, they are quite resistant to freezing, and may survive in some foods for a number of years (26).

Research with both naturally and artificially contaminated eggs has demonstrated that, in the great majority of eggs, there would appear to be little growth of *S.* Enteritidis until they have been stored at 20 °C for approximately 21 days. The principal site of contamination in eggs seems to be either the outside of the vitelline (yolk) membrane or the albumen surrounding it. *Salmonella* Enteritidis is unable to grow until storage-related permeability changes have taken place to the vitelline membrane, which allow the organism to invade yolk contents. Once this occurs, large populations of salmonellas will be achieved in both the yolk and the albumen. Membrane changes take place more quickly at higher temperatures and at high humidity. Under simulated kitchen conditions, eggs were able to support the rapid growth of *S.* Enteritidis within a few days. Such growth does not become obvious until the population exceeds 10^9 cells/egg (19,27).

Heat resistance

Salmonella is not a spore-forming organism. It is not, therefore, a heat-resistant organism; pasteurisation and equivalent heat treatments will destroy the organism under normal circumstances. $D_{60°C}$ values normally range from about 1-10 min, with a z-value of 4-5 °C. However, high fat or low moisture (low a_w) will reduce the effectiveness of heat treatments, and appropriate heat treatments must be determined experimentally for low-a_w foods. For example, it is not possible to apply a sufficient heat process to decontaminate chocolate without causing an unacceptable change in the product. *S.* Enteritidis, *S.* Typhimurium and *S.* Agona were found to have a survival rate of 0.1% after heating to 70-90 °C for 20-30 min (28). Strains vary in their ability to withstand heating; *S.* Senftenberg 775W is about 10-20 times more heat-resistant than the average strain of *Salmonella* at high a_w (26). Recent research has shown that exposure to a sub-lethal heat shock may also cause cells to show an increase in heat resistance (29).

pH

Salmonella has a pH range for growth of pH 3.8-9.5, under otherwise ideal conditions, although most serotypes will not grow below 4.5 (26). The acid used influences the minimum pH for growth to occur. For example, acetic acid (e.g. when vinegar is used in mayonnaise, rather than citric acid) will generally limit the pH minimum to about pH 5. Some death will occur at pH values of less than about 4.0, depending on the type of acid and temperature, but recent evidence has shown that an acid shock can trigger a complex response in cells allowing them to survive in such environments (30). The optimal pH for *Salmonella* growth is between 6.5 and 7.5.

Water activity/salt

Where all other conditions are favourable, *Salmonella* has the potential to grow at a_w levels as low as 0.945, or possibly 0.93, depending on serotype, substrate, temperature and pH. Salmonellae are quite resistant to drying as demonstrated by an *S.* Agona outbreak associated with toasted oat cereal that occurred in the US in 1998 (31). Survival for many years in low moisture, high fat foods, such as chocolate, has been reported (32).

The growth of *Salmonella* is generally inhibited by the presence of 3-4% sodium chloride, although salt tolerance increases with increasing temperature (33).

Atmosphere

Salmonellae can grow both aerobically and anaerobically, although growth can be inhibited by an Eh potential (oxidation - reduction or redox) below -30 mV (17).

Irradiation

Salmonella is more resistant to irradiation than other Gram-negative pathogens, and a D-value of 0.6 kGy has been reported (34). Decontamination doses for poultry of about 3 kGy are reported to give approximately a 3-log reduction in numbers of *Salmonella* (35).

The lethal effect of irradiation is enhanced by the presence of oxygen, and USDA Food Safety and Inspection Service regulations state that the packaging used for irradiated poultry must be air-permeable (36).

Summary of Control of *Salmonella* in Foods

All sectors in the food industry, as well as consumers, need to be involved in the control of *Salmonella* at all points in the food chain. This demands the control of animal feed, hygiene during livestock production and processing, through to the prevention of cross-contamination from raw food (especially raw poultry) to foods that are to receive no further heat treatment - during retail and catering, and in the home.

It is also important to prevent the contamination of fresh produce during cultivation, harvesting, and processing, by implementing good agricultural practice, such as control of irrigation water quality, excluding livestock and wildlife from growing areas, and good hygiene during washing and packing.

Proper attention must be paid to the control of temperature to prevent the growth of *Salmonella* during food storage. The reformulation of foods - for example, by reducing the salt or acidity levels - can also give rise to conditions suitable for survival or growth.

The HACCP approach should be applied to ensure that proper controls are in place to destroy or prevent contamination with *Salmonella* during food manufacturing, processing and handling. It is particularly important to ensure that salmonellae are not present in ready-to-eat foods in view of the potentially low infective dose.

Although the prevalence of *Salmonella* in eggs has been successfully reduced by the vaccination of laying flocks and other measures, the UK Egg

Products Regulations 1993 require egg products sold or used in the preparation of food to comply with specified requirements as to heat treatment, sampling, storage and transport. The UK Department of Health advises all consumers to avoid eating raw eggs, and vulnerable individuals only to eat eggs that are thoroughly cooked. The ACMSF also recommended that eggs should be consumed within three weeks of date of lay and should be labelled with a "use by" date. Once purchased, either by the caterer or the consumer, they should be refrigerated. The use of pasteurised, rather than shell eggs, is also encouraged (19).

Bibliography

References

1. Le Minor L., Popoff M.Y. Request for an opinion. Designation of *Salmonella enterica* sp. Nov., nom. Rev., as the type and only species of the genus *Salmonella. International Journal of Systematic Bacteriology,* 1987, 37, 465-8.

2. Threlfall J., Ward L., Old D. Changing the nomenclature of *Salmonella. Communicable Disease and Public Health,* 1999 (September), 2 (3), 156-7.

3. Sharp J.C.M. Salmonellosis. *British Food Journal,* 1990, 92 (4), 2+6-12.

4. Threlfall E.J., Ward L.R., Rowe B. Increasing incidence of resistance to trimethoprin and ciprofloxacin in epidemic *Salmonella typhimurium* DT104 in England and Wales. *Eurosurveillance,* 1997 (November), 2 (11), 81-4.

5. Walker R.A., Lawson A.J., Lindsay E.A., Ward L.R., Wright P.A., Bolton F.J., Wareing D.R., Corkish J.D., Davies R.H., Threlfall E.J. Decreased susceptibility to ciprofloxacin in outbreak-associated multiresistant *Salmonella typhimurium* DT104. *Veterinary Record,* 2000 (September), 147 (14), 395-6.

6 World Health Organization. Fact sheet No 139, Revised April 2005.

7. D'Aoust J.-Y. *Salmonella* and the international food trade. International *Journal of Food Microbiology,* 1994, 24 (1/2), 11-31.

8. Greenwood M.H., Hooper W.L. Chocolate bars contaminated with *Salmonella napoli*: an infectivity study. *British Medical Journal,* 1983, 286 (6375), 1394.

9. Vought K.J., Tatini S.R. *Salmonella enteritidis* contamination of ice cream associated with a 1994 multistate outbreak. *Journal of Food Protection,* 1998 (January), 61 (1), 5-10.

10. Cowden J.M., O'Mahony M., Bartlett C.L.R., Rana B., Smyth B., Lynch D., Tillett H., Ward L., Roberts D., Gilbert R.J., Baird-Parker A.C., Kilsby D.C. A national outbreak of *Salmonella typhimurium* DT124 caused by contaminated salami sticks. *Epidemiology and Infection,* 1989, 103 (2), 219-25.

11. Ryan C.A., Nickels M.K., Hargrett-Bean N.T. Massive outbreak of antimicrobial-resistant salmonellosis traced to pasteurised milk. *Journal of the American Medical Association,* 1987, 258 (22), 3269-74.

12. Rowe B., Hutchinson D.N., Gilbert R.J., Hales B.H., Begg N.T., Dawkins H.C., Jacob M., Rae F.A., Jepson M. *Salmonella ealing* infections associated with consumption of infant dried milk. *Lancet,* 1987 (8564), 900-3.

13. Hennessy T.W., Hedberg C.W., Slutsker L., White K.E., Besser-Wiek J.M., Moen M.E., Feldman J., Coleman W.W., Edmonson L.M., MacDonald K.L., Osterholm M.T. A national outbreak of *Salmonella enteritidis* infections from ice cream. *New England Journal of Medicine,* 1996, 334 (20), 1281-6.

14 Kivi M, van Pelt W, Notermans D, van de Giessen A, Wannet W, Bosman A. Large outbreak of *Salmonella* Typhimurium DT104, the Netherlands, September–November 2005. Euro Surveill 2005;10(12):E051201.1. Available from: http://www.eurosurveillance.org/ew/2005/051201.asp#1.

15. Anon. Up and coming 'new types' of *Salmonella* in England and Wales. *CDR Weekly,* 2002 (July 11), 12 (28).

16. Vugia D., Hadler J., Blake P., Blythe D., Smith K., Morse D., Cieslak P., Jones T., Shillam P., Chen D.W., Garthright B., Charles L., Molbak K., Angulo F., Griffin P., Tauxe R. Preliminary FoodNet data on the incidence of foodborne illnesses - selected sites, United States, 2001. *Morbidity and Mortality Weekly Report (MMWR),* 2002, (April 12), 51 (14), 325-9.

17. Doyle M.P., Cliver D.O. *Salmonella.* Foodborne Diseases. Ed. Cliver D.O. London. *Academic Press,* 1990, 185-204.

18. Henson S. Estimating the incidence of foodborne *Salmonella* and the effectiveness of alternative control measures using the Delphi method. *International Journal of Food Microbiology,* 1997 (April 15), 35 (3), 195-204.

19. Advisory Committee on the Microbiological Safety of Food. Report on *Salmonella* in eggs. London. *HMSO,* 1993.

20. Anon. Less *Salmonella* in retail chicken. *Veterinary Record,* 2001 (September), 149 (11), 314-5.

21. Advisory Committee on the Microbiological Safety of Food. Report on poultry meat. Annex D: survey of *Salmonella* contamination in UK-produced raw chicken on retail sale. Ed. Advisory Committee on the Microbiological Safety of Food. London. *HMSO,* 1996, 110-30.

22. Dominguez C., Gomez I., Zumalacarregui J. Prevalence of *Salmonella* and *Campylobacter* in retail chicken meat in Spain. *International Journal of Food Microbiology,* 2002 (January 30), 72 (1-2), 165-8.

23. D'Aoust J.-Y., Maurer J., Bailey J.S. *Salmonella* species. Food microbiology: fundamentals and frontiers. Eds Doyle M.P., Beuchat L.R., Montville T.J. 2nd edition. Washington DC. ASM Press, 2001, 141-78.

24. Heinitz M.L., Ruble R.D., Wagner D.E., Tatini S.R. Incidence of *Salmonella* in fish and seafood. *Journal of Food Protection,* 2000 (May), 63 (5), 579-92.

25. Kim *et al.* Effect of time and temperature on growth of *Salmonella enteritidis* in experimentally inoculated eggs. *Avian Disease,* 1989, 33, 735-42.

26. International Commission on Microbiological Specifications for Foods Microorganisms in Foods. Salmonellae, in *Microorganisms in Foods, Volume 5: Microbiological Specifications of Food Pathogens.* Ed. International Commission on Microbiological Specifications for Foods. London. Blackie, 1996, 217-64

27. Humphrey T.J. Public health aspects of contamination of eggs and egg products with salmonellas. London. SCI, 1995, *SCI Lecture Papers* No. 0065.

28 Shachar D., Yaron S. Heat Tolerance of *Salmonella* entera Serovars Agona, Enteritidis and Typhimurium in Peanut Butter. *Journal of Food Protection,* Nov 2006, 69 (11) 2687-92.

29. Xavier K.J., Ingham S.C. Increased D-values for *Salmonella enteritidis* following heat shock. *Journal of Food Protection,* 1997, 60 (2), 181-4.

30. D'Aoust J.-Y. *Salmonella* species, in *Food Microbiology: Fundamentals and Frontiers.* Eds Doyle M.P., Beuchat L.R., Montville T.J. Washington DC. ASM Press, 1997, 129-58.

31. Anon. Multistate outbreak of *Salmonella* serotype Agona infections linked to toasted oat cereal - United States, April-May, 1998. *Morbidity and Mortality Weekly Report,* 1998 (June 12), 47 (22), 462-4.

32. D Aoust J.-Y. *Salmonella* and the chocolate industry. A review. *Journal of Food Protection,* 1977, 40 (10), 718-27 (140 ref.).

33. D'Aoust J.-Y. *Salmonella*, in *Foodborne Bacterial Pathogens.* Ed. Doyle M.P. New York. Marcel Dekker, 327-445.

34. Mendonca A.F. Inactivation by irradiation, in *Control of foodborne microorganisms*. EdsJuneja V.K., Sofos J.N. New York. Marcel Dekker, 2002, 75-103.

35. Kampelmacher E.H. Food irradiation- a new technology for preserving foods and keeping them hygienically safe. *Fleischwirtschaft,* 1983, 63 (11), 1677-86 (5).

36. Title 21 of the Code of Federal Regulations. Part 179.26.

Further reading

Food Outbreak response published March 2006 http://www.cieh.org/ehp/food_safety/articles/food_outbreak_response.htm.

Bell C., Kyriakides A. *Salmonella*, in *Foodborne pathogens: hazards, risk analysis and control.* Eds Blackburn C. de W., McClure P.J. Cambridge. Woodhead Publishing Ltd, 2002, 307-35.

D'Aoust J.-Y., Maurer J., Bailey J.S. *Salmonella* species, in *Food microbiology: fundamentals and frontiers.* Eds Doyle M.P., Beuchat L.R., Montville T.J. 2nd edition. Washington DC. ASM Press, 2001, 141-78.

D'Aoust J.-Y. *Salmonella*, in *Guide to foodborne pathogens.* Eds Labbe R.G., Garcia S. New York. Wiley, 2001, 163-91.

Bell C., Kyriakides A. *Salmonella*: a practical approach to the organism and its control in foods. Oxford. Blackwell Science, 2001.

D'Aoust J.-Y. *Salmonella*, in *The microbiological safety and quality of food, volume 2.* Eds Lund B.M., Baird-Parker T.C., Gould G.W. Gaithersburg. Aspen Publishers, 2000, 1233-99.

Jay L.S., Grau F.H., Smith K., Lightfoot D., Murray C., Davey G.R. *Salmonella,* in *Foodborne Microorganisms of Public Health Significance.* Ed. Australian Institute of Food Science and Technology. 5th Edn North Sydney: *AIFST,* 1997, 169-230.

Various authors. *Salmonella* Food Associated Pathogens: Proceedings of a Symposium. Uppsala, May 1996 Uppsala. International Union of Food Science and Technology. SLU, 1996, 152-68.

Steinhart C.E., Doyle M.E., Cochrane B.A. Foodborne bacterial intoxications and infections. *Salmonella* Food Research Institute, Food Safety 1996. Food Research Institute. New York. Marcel Dekker, 1996, 414-37.

Foster J.W., Spector M.P. How *Salmonella* survive against the odds. Annual Review of Microbiology, Volume 49 Ed. Ornston L.N. Palo Alto. Annual Reviews Inc., 1995, 145-74.

Methods for detection

Ryser E.T., Donnelly C.W., Andrews W.H., Flowers R.S., Silliker J., Bailey J.S., Lampel K.A. *Listeria*, *Salmonella* and *Shigella*. Compendium of methods for the microbiological examination of foods. American Public Health Association, Downes F.P., Ito K. 4th edition, Washington DC. APHA, 2001, 343-85.

Andrews W.H. Microbiological methods. Subchapter 9. *Salmonella* (parts 1-3). Official methods of analysis of AOAC International, volume 1: agricultural chemicals, contaminants, drugs. AOAC International, Horwitz W. 7th edition. Gaithersburg. AOAC International, 2000, Chapter 17, 78-137.

Public Health Laboratory Service. Detection of *Salmonella* species. PHLS standard methods for food products. Public Health Laboratory Service, London. PHLS, 1999, 8.

British Standards Institution. Microbiology of food and animal feeding stuffs - horizontal method for the detection of *Salmonella*. BS EN 12824:1998 ISO 6579:1993 (modified), 1998.

Afflu L., Gyles C.L. A comparison of procedures involving Single Step *Salmonella*, 1-2 Test, and Modified Semisolid Rappaport-Vassiliadis medium for detection of *Salmonella* in ground beef. *International Journal of Food Microbiology,* 1997 (July 22), 37 (2-3), 241-4.

Hanai K., Satake M., Nakanishi H., Venkateswaran K. Comparison of commercially available kits with standard methods for detection of *Salmonella* strains in foods. *Applied and Environmental Microbiology,* 1997, 63 (2), 775-8.

Davies R.H. Principles and practice of *Salmonella* isolation, in *Factors Affecting the Microbial Quality of Meat, Volume 4: Microbial Methods for the Meat Industry.* Eds Commission of the European Communities, Hinton M.H., Rowlings C. Bristol. University of Bristol Press, 1996, 15-26.

Andrews W.H., June G.A., Sherrod P., Hammack T.S., Amaguana R.M. *Salmonella,* in Bacteriological Analytical Manual. Ed. Food and Drug Administration. 8th Edn. Gaithersburg. AOAC International, 1995.

Milk and milk products. Detection of *Salmonella*. IDF Standard 93B: 1995. *International Dairy Federation*, 1995.

Busse M. Media for *Salmonella.* Culture Media for Food Microbiology Eds Corry J.E.L., Curtis G.D.W., Baird R.M. Amsterdam. Elsevier, 1995, 187-2091.

Blackburn C. de W. Rapid and alternative methods for the detection of salmonellas in foods. *Journal of Applied Bacteriology,* 1993, 75 (3), 199-214.

Flowers R.S., D'Aoust J.-Y., Andrews W.H., Bailey J.S. *Salmonella*, in *Compendium of Methods for the Microbiological Examination of Foods*. Ed. American Public Health Association, Vanderzant C., Splittstoeser D.F. Washington DC. APHA, 1992, 371-422.

STAPHYLOCOCCUS AUREUS

Although there are records of the illness as early as 1830 the organism was not recognised as a foodborne pathogen until sometime between 1878 and 1880.

Staphylococcal food poisoning is generally caused by the ingestion of a toxin pre-formed in contaminated food. Although the illness produced is not considered to be particularly serious, it is quite common on a worldwide basis and occurs either sporadically or in significant outbreaks. The principal species involved in cases of food poisoning is *Staphylococcus aureus*.

The Organism *Staphylococcus aureus*

Staphylococcus aureus is a Gram-positive, catalase-positive, oxidase-negative coccus, belonging to the family Micrococcaceae, which is capable of producing heat-stable toxins (enterotoxins) in food. Cells form characteristic clumps resembling bunches of grapes. It can grow either aerobically or anaerobically (1,2,3).

In addition to *Staph. aureus* there are at least 29 other species belonging to the genus *Staphylococcus* (4). However, *Staph. aureus* is the species that is of main concern to food microbiologists. Between 40 and 50% of staphylococcal isolates from healthy humans are capable of producing enterotoxins. *Staph. aureus* can also produce coagulase and thermonuclease (3,5).

Other coagulase or thermonuclease-positive staphylococci (certain strains of *Staph. hyicus* and *Staph. intermedius*) can produce enterotoxin (5,6). In a food-related outbreak in the US during 1991, *Staph. intermedius* was implicated as the aetiological agent. In this outbreak there were over 265 cases associated with the consumption of a 'butter-blend spread' (6,7). It has also been demonstrated more recently that some coagulase and thermonuclease-negative staphylococci are also capable of producing enterotoxin (5,8).

It is the enterotoxins produced by the staphylococci that are known to be the emetic cause of food poisoning. These enterotoxins are heat-stable, hygroscopic, water-soluble proteins. There are 11 distinct staphylococcal enterotoxins: A, B, C1, C2, C3, D, E, G, H and I. Most food-poisoning strains are found to produce A, with D being the second most frequent. The fewest number of outbreaks is associated with E (2,5).

Staph. aureus Food Poisoning

When consumption of food has resulted in staphylococcal food poisoning, at least two errors have been made in its manufacture. The first is contamination with the organism, and the second is that conditions favourable for growth and toxin production occurred at some stage during production or storage (9).

The organism can reach numbers to produce sufficient enterotoxin to cause illness, but then may be killed in further processing or die out under the conditions of storage.

Incubation time

The usual symptoms develop abruptly within 1-7 h. The length of incubation time, and severity of illness, depends on the amount of enterotoxin ingested (10).

Symptoms

Symptoms normally include nausea and vomiting, retching, with occasional abdominal cramping, and diarrhoea. Headache and muscle cramping and sometimes, later, dehydration, may occur in more severe cases (2).

Mortality

Recovery is normally rapid, within a few hours or a day or so. However, deaths - amongst children and the elderly - have occurred, but rarely (2,3).

Infective dose

As already mentioned, no live organisms of *Staph. aureus* need to be ingested. However, *Staph. aureus* generally needs to grow to levels of 10^5-10^6/g food for sufficient toxin to be produced. Toxin dose of 1 μg or less is needed to cause food poisoning (6,11). In a well documented outbreak associated with chocolate milk, the quantity of enterotoxin was estimated at 144 ng +/- 50 (2,3,12).

Foods involved

Foods involved in *Staph. aureus* food poisoning are commonly those that have been physically handled and temperature-abused prior to consumption. Thus, cooked meat and poultry products (where the number of competing micro-organisms has been reduced through cooking), milk and cream, custard- or cream-filled pastries, butter, salads containing potato, egg or prawns, and other moist foods, such as cold sweets, are susceptible to the growth of *Staph. aureus,* and have been involved in *Staph. aureus* food poisoning (1,3,11).

Salted meat, such as ham, has also quite commonly been implicated. This meat product is partly preserved by the presence of salt, which inhibits spoilage bacteria. *Staph. aureus*, however, is able to tolerate relatively high levels of salt concentration, so is able to grow in the absence of competitor organisms if temperature-abuse occurs (6,13).

Other foods sometimes implicated include cheese - especially where there has been a starter culture failure (leading to insufficient/too slow acid production).

During the period 1969-90, poultry and meat products were implicated in 75% of incidents of staphylococcal food poisoning reported in the UK, with ham and chicken most commonly implicated. The most frequently reported place where food poisoning occurred was in the home, followed by restaurants and shops (14).

In the UK, during 1992-93, there were 8 outbreaks due to staphylococcal food poisoning, and the foods implicated were chicken, corned beef, pork, ham/bacon, chicken salad and cheese flan (15).

In Japan, rice balls is the major item involved in staphylococcal food poisoning (3).

Incidence of *Staph. aureus* Food Poisoning

The nature of the food poisoning caused by this organism, i.e. relatively mild symptoms and rapid recovery time, means that it is not commonly reported. Estimates in the US suggest that only between 1and 5% of cases are reported (16). The incidence of the disease in different countries varies according to geography and eating habits. In the US during 1983-87, 47 (8%) outbreaks (3,181 cases) of the total 600 bacterial outbreaks were due to *Staph. aureus*, whereas in England and Wales for the same time period 54 (2%) outbreaks out of a total of 2,815 bacterial outbreaks were recorded (6,17). During the period 1969-90, there was an annual average of 10-15 incidents of *Staph. aureus* food poisoning reported in England and Wales. Since 1980, the number of reported cases in the UK has not exceeded 189 per annum (2). Half of these incidents occurred during the summer months, and hot summers saw a doubling in the average annual incidence (14). Figures for 1996 showed that in England and Wales, five general outbreaks and a total of 150 individual cases were reported. A recent case in Japan in the year 2000 involved 13,420 individuals (18).

Sources

Humans

In contrast to other major types of food poisoning, humans (viz. food handlers) play a major role in the transmission of the causative organism.

Humans are an important source of *Staph. aureus* - the organism can be carried in the hair, throat and nose (it can be associated with 'sore throats' or colds), on the skin, and transferred to foods, especially via the hands. It has been reported that about 40% of healthy people are nasopharyngeal carriers of *Staph. aureus* (19).

Infected cuts and sores on hands and faces can also serve as a source of the bacterium for foods.

Animals and environment

Animals can act as a source of *Staph. aureus* - through raw milk in the case of mastitic cows (mastitis can be caused by *Staph. aureus*), or through raw meat (particularly pig meat), which is commonly contaminated with *Staph. aureus* (5). *Staph. aureus* is also quite resistant to drying, and may persist for long periods on processing equipment such as meat slicers, or in dust within air handling equipment.

Staphylococci can be found in air, dust, water, and human and animal waste.

Foods

A high proportion of raw meats, including beef carcases, ground beef, salami, pork sausage, and pork carcases are contaminated with staphylococci. These organisms are also often present on raw poultry, and various types of seafood (7).

In general, any foods that are of animal origin or that have been physically handled and not subsequently given a bactericidal treatment may carry the risk of being contaminated with *Staph. aureus* (5).

Growth/Survival Characteristics of the Organism in Foods

The growth of *Staph. aureus* is more limited under anaerobic than under aerobic conditions. The physicochemical limits for toxin production are also narrower than for growth.

Temperature

Under otherwise ideal conditions *Staph. aureus* can grow within the temperature range 7-48.5 °C (18), with an optimum of 30-37 °C (18). Enterotoxin production has been reported in the range 10-45 °C, with an optimum of 35-40 °C (3), and is inhibited more by lower temperatures of food storage. Longer storage at lower (10 °C), but still abusing, temperatures can result in sufficient toxin production to cause illness (3).

Freezing and thawing have little effect on *Staph. aureus* viability, but may cause some cell damage (1).

Heat resistance

Heat resistance depends very much on the food type in which the organism is being heated (conditions relating to pH, fat content, water activity, etc.). As is the case with other bacteria, stressed cells can also be less tolerant of heating.

Under most circumstances, however, the organism is heat-sensitive and will be destroyed by pasteurisation (2,13).

Examples of reported D-values

In:

Phosphate buffer (pH 7): $D_{56\,°C}$ 1-2 min (z-value 8-10 °C)

Meat: $D_{60\,°C}$ 2-20 min (depending on a_w)

Milk: $D_{60\,°C}$ 1-6 min (z-value 7-9 °C)

Egg pasta: $D_{60\,°C}$ 3-40 min (depending on a_w)

Oil: $D_{120\,°C}$ 3-6 min.

The enterotoxins are quite heat-resistant. Destruction depends on temperature, pH and toxin type. In general, heating at 100 °C for at least 30 min may be required to destroy unpurified toxin (3,13). Enterotoxins may potentially survive thermal processes applied in the sterilisation of low-acid foods (4).

pH

The pH at which a staphylococcal strain will grow is dependent on the type of acid (acetic acid is more effective at destroying *Staph. aureus* than citric acid), water activity and temperature (sensitivity to acid increases with increasing temperature). Most strains of staphylococci can grow within the pH range 4.2-9.3 (optimum 7.0-7.5) (3,18) under otherwise ideal conditions.

The pH range supporting enterotoxin production is narrower (pH 5.2-9.0) than the one supporting growth, namely pH 4.5-9.3 (18). The optimal pH for the production of enterotoxin is between 7.0-7.5, depending on strain and type of toxin (3).

Water activity/salt

Staph. aureus is unusual amongst food-poisoning organisms in its ability to tolerate low water activities. It can grow over the a_w range 0.83 - >0.99 aerobically under otherwise optimal conditions. However, a_w of 0.86 is the generally recognised minimum in foods (5).

The water activity level supporting toxin production is dependent on temperature, pH, atmosphere, and strain and solute. In general, the range is 0.87 - >0.99.

Staphylococci are more resistant to salt present in foods than other organisms. In general, *Staph. aureus* can grow in 7-10% NaCl, but certain strains can grow in 20%. An effect of increasing salt concentration is to raise the minimum pH of growth. Enterotoxin production has been demonstrated in the presence of about 10% NaCl. The yield of toxin decreased with increasing salt concentration dependent on pH and temperature (5,20).

Atmosphere

Although *Staph. aureus* is capable of anaerobic growth, it grows best in the presence of air.

There is little or no toxin produced under anaerobic conditions, especially vacuum-packed foods (20).

Summary of Control of *Staph. aureus* in Foods

Control against *Staph. aureus* food poisoning is essentially dependent on two factors:

1) minimisation of physical handling
2) control of temperature

Every effort should be made to prevent the contamination of foods - especially processed/cooked foods - through the poor personal hygiene of handlers. Where handling is unavoidable scrupulous attention should be paid to hand hygiene. Ideally, people with colds, or with infected cuts/sores on their hands, should be excluded from directly handling processed or ready-to-eat foods.

The control of temperature is essential at all times. Susceptible foods must be kept either sufficiently cold (≤4 °C) or hot (≥63 °C) to prevent the growth of *Staph. aureus*. Ideally, if food is to be stored it should be in shallow, covered containers.

Note:

Decisions about the significance of detecting staphylococci in foods need to be made with care. The mere presence of the organism in foods, especially raw foods, such as raw milk or meat, need not always be cause for concern. The organism needs to grow to large numbers in order to form toxin and present a hazard. This requires the abuse of temperature and the lack of competitor organisms (*Staph. aureus* does not compete well against spoilage bacteria unless they outnumber competitors initially).

Nevertheless, if foods containing *Staph. aureus* are used as ingredients in other foods that may offer a favourable environment for the growth of *Staph. aureus*, this may represent a hazard. Furthermore, it must always be remembered that staphylococcal enterotoxins are very heat-resistant, and will withstand the usual cooking/reheating practices.

Bibliography

References

1. Reed G.H. Foodborne illness (Part 1): Staphylococcal ("Staph") food poisoning. *Dairy, Food and Environmental Sanitation,* 1993, 13 (11), 642.
2. Sutherland J., Varnam A. Enterotoxin-producing *Staphylococcus*, *Shigella*, *Yersinia*, *Vibrio*, *Aeromonas* and *Plesiomonas,* in *Foodborne Pathogens Hazards, risk analysis and control*. Eds Blackburn C. de W., McClure P.J. Cambridge. Woodhead Publishing Ltd., 2002, 385-90.
3. Bergdoll M.S, Lee Wong A.C. Staphylococcal intoxications, in *Foodborne infections and intoxications*. Eds Riemann H.P., Cliver D.O. London. Academic Press, 2005, 523-62.
4. *Staphylococcus aureus*. Microorganisms in Foods, Vol. 5. Microbiological Specifications of Food Pathogens Ed. International Commission on Microbiological Specifications for Foods. London. Blackie, 1996, 513.

5. Jay J.M., Loessner M.J., Golden D.A. Staphylococcal gastroenteritis, in *Modern food microbiology.* Eds Jay J.M., Loessner M.J., Golden D.A. New York. Springer Science, 2005, 545-66.

6. Bennett R.W., Monday S.R. *Staphylococcus aureus*, in International Handbook of Foodborne Pathogens. Eds Miliotis M.D., Bier J.W. New York. Marcel Dekke, .2003, 41-60.

7. Khambaty F.M., Bennett R.W., Shah D.B. Application of pulsed-field gel electrophoresis to the epidemiological characterisation of *Staphylococcus* intermedius implicated in a food-related outbreak. *Epidemiology and Infection*, 1994, 113 (1), 75-81.

8. Bennett R.W. Atypical toxigenic *Staphylococcus* and non-*Staphylococcus aureus* species on the horizon? An update. *Journal of Food Protection,* 1996, 59 (10), 1123-6.

9. Mossel D.A.A., van Netten P. *Staphylococcus aureus* and related staphylococci in foods: ecology, proliferation, toxinogenesis, control and monitoring. *Journal of Applied Bacteriology,* 1990, 69, supplement 'Staphylococci', 123S-45S, 19.

10. A working party of the PHLS *Salmonella* Committee. The prevention of human transmission of gastrointestinal infections, infestations and bacterial intoxications. *CDR Review*, 1995, 5 (11), R157-72.

11. Martin S.E., Myers E.R. *Staphylococcus aureus,* in *Foodborne Disease Handbook, Vol. 1: Diseases Caused by Bacteria.* Eds Hui Y.H., Gorham J.R., Murrell K.D., Cliver D.O. New York. Marcel Dekker, 1994, 345-94.

12. Evanson M.L., Ward Hinds M., Berstein R.S., Bergdoll M.S. Estimation of human dose of staphylococcal enterotoxin A from a large outbreak of staphylococcal food poisoning involving chocolate milk. *International Journal of Food Microbiology,* 1988, 7(4) 311-6.

13. Stewart G.C. *Staphylococcus aureus,* in *Foodborne pathogens: microbiology and molecular biology.* Eds Fratamico P.M., Bhunia A.K., Smith J.L. Wymondham. Caister Academic Press, 2005, 273-84.

14. Wieneke A.A., Roberts D., Gilbert R.J. Staphylococcal food poisoning in the United Kingdom, 1969-90. *Epidemiology and Infection,* 1993, 110 (3), 519-31.

15. Cowden J.M., Wall P.G., Adak G., Evans H., Le Baigue S., Ross D. Outbreaks of foodborne infectious intestinal disease in England and Wales: 1992 and 1993. *CDR Review,* 1995, 5 (8), R109-17.

16. Jablonski L.M., Bohach G.A. *Staphylococcus aureus*, in *Food Microbiology: Fundamentals and Frontiers*. Eds Doyle M.P., Beuchat L.R., Monteville T.J. Washington DC. ASM Press, 1997, 353-75.

17. Bean N.H., Griffin P.M., Goulding J.S., Ivey C.B. Foodborne disease outbreaks, 5-year summary, 1983-87. *Morbidity and Mortality Weekly Report,* 1990, 39 (SS-1), 15-57.

18. Gustafson J., Wilkinson. *Staphylococcus aureus* as a food pathogen: staphylococcal enterotoxins and stress response systems, in *Understanding pathogen behaviour Virulence, stress response and resistance.* Ed. Griffiths M. Cambridge. Woodhead Publishing Ltd., 2005, 331-57.

19. Management of Outbreaks of Foodborne Illness. *Staphylococcus aureus* enterotoxin. Department of Health. Heywood. Department of Health, 1994, 135.

20. Bergdoll M.S. Staphylococcal Food Poisoning, in *Foodborne Disease.* Ed. Cliver D.O. London. Academic Press, 1990, 85-106.

Further reading

Soejima T., Nagao E., Yano Y., Yamagata H., Kagi H., Shinagawa K. Risk evaluation for staphylococcal food poisoning in processed milk produced with skim milk powder. *International Journal of Food Microbiology,* 2007, (April 1), 115 (1), 29-34.

Colombari V., Mayer M.D.B., Laicini Z.M., Mamizuka E., Franco B.D.G.M., Destro M.T., Landgraf M. Foodborne outbreak caused by *Staphylococcus aureus*: phenotypic and genotypic characterization of strains of food and human sources. *Journal of Food Protection,* 2007, (February), 70 (2), 489-93.

Bania J., Dabrowska A., Bystron J., Korzekwa K., Chrzanowska J., Molenda J. Distribution of newly described enterotoxin-like genes in *Staphylococcus aureus* from food. *International Journal of Food Microbiology,* 2006, (April 15), 108 (1), 36-41.

Engel R.A., Fanslau M.A., Schoeller E.L., Searls G., Buege D.R., Zhu J. Fate of *Staphylococcus aureus* on vacuum-packaged ready-to-eat meat products stored at 21 °C. *Journal of Food Protection,* 2005 (September), 68 (9), 1911-15.

Jablonski L.M. Bohach G.A. *Staphylococcus aureus*, in *Food Microbiology: Fundamentals and Frontiers. 2nd edition.* Eds Doyle M.P., Beuchat L.R., Montville T.J. Washington DC. ASM Press, 2001, 411-34.

Baird-Parker T.C. *Staphylococcus aureus*, in *The microbiological safety and quality of food, volume 2.* Eds Lund B.M., Baird-Parker T.C., Gould G.W. Gaithersburg. Aspen Publishers, 2000, 1317-35.

Martin S.E., Myers E.R., Iandolo J.J. *Staphylococcus aureus*, in *Foodborne Disease Handbook, volume 1: bacterial pathogens. 2nd edition.* Eds Hui Y.H., Pierson M.D., Gorham J.R. New York. Marcel Dekker, 2000, 345-81

International Commission on Microbiological Specifications for Foods. *Staphylococcus aureus*. Microorganisms in Foods, vol 5. Microbiological Specifications of Food Pathogens. Ed. International Commission on Microbiological Specifications for Foods. London. Blackie, 1996, 299-333.

Asperger H. *Staphylococcus aureus*. The Significance of Pathogenic Microorganisms in Raw Milk. Ed. International Dairy Federation. Brussels. IDF, 1994, 24-42.

Baird-Parker A.C. The staphylococci: an introduction. *Journal of Applied Bacteriology,* 1990, 69, supplement 'Staphylococci', 1S-8S, 19.

Methods of detection

Goto M., Takahashi H., Segawa Y., Hayashidani H., Takatori Y. Real-time PCR method for quantification of *Staphylococcus aureus* in milk. *Journal of Food Protection,* 2007 (January), 70 (1), 90-96.

Tomasino S.F., Fiumara R.M., Cottrill M.P. Enumeration procedure for monitoring test microbe populations on inoculated carriers in AOAC use-dilution methods. *Journal of AOAC International,* 2006, (November-December), 89 (6), 1629-34.

Rajkovic A., el Moualij B., Uyttendaele M., Brolet P., Zorzi W., Heinen E., Foubert E., Bebevere J. Immunoquantitative real-time PCR for detection and quantification of *Staphylococcus aureus* enterotoxin B in foods. *Applied and Environmental Microbiology,* 2006, (October), 72 (10), 6593-9.

Medina M.B. Development of a fluorescent latex microparticle immunoassay for the detection of staphylococcal enterotoxin B (SEB). *Journal of Agricultural and Food Chemistry,* 2006, (July 12), 54 (14), 4937-42.

Lamprell H., Mazerolles G., Kodjo A., Chamba J.F., Noel Y., Beuvier E. Discrimination of *Staphylococcus aureus* strains from different species of Staphylococcus using Fourier transform infrared (FTIR) spectroscopy. *International Journal of Food Microbiology,* 2006, (April 15), 108 (1), 125-9.

Alarcon B., Vicedo B., Aznar R. PCR-based procedures for detection and quantification of *Staphylococcus aureus* and their application in food. *Journal of Applied Microbiology,* 2006, (February), 100 (2), 352-64.

British Standards Institution. Microbiology of food and animal feeding stuffs. Horizontal method for the enumeration of coagulase-positive staphylococci (*Staphylococcus aureus* and other species). Detections and MPN technique for low numbers. BS EN ISO 6888-3:2006. 2006.

Gandra E.A., Silva J.A., de Macedo M.R.P., de Araujo M.R., Mata M.M., da Silva W.P. Differentiation between *Staphylococcus aureus*, *S. intermedius* and *S. hyicus* using phenotypical tests and PCR. *Alimentos e Nutricao (Brazilian Journal of Food and Nutrition)*. 2005, (April-June), 16 (2), 99-103.

Delbes C, Montel M.C. Design and application of a *Staphylococcus*-specific single strand conformation polymorphism-PCR analysis to monitor *Staphylococcus* populations diversity and dynamics during production of raw milk cheese. *Letters in Applied Microbiology,* 2005, 41 (2), 169-74.

Pinyo B, Chenoll E, Aznar R. Identification and typing of food-borne *Staphylococcus aureus* by PCR-based techniques. *Systematic and Applied Microbiology,* 2005, (June), 28 (4), 340-52.

Villard L, Lamprell H, Borges E, Maurin F, Noel Y., Beuvier E, Chamba J.F, Kodjo A. Enterotoxin D producing strains of *Staphylococcus aureus* are typeable by pulsed-field gel electrophoresis (PFGE). *Food Microbiology,* 2005, (April-June), 22 (2-3), 261-5.

Lancette G.A., Bennett R.W. *Staphylococcus aureus* and staphylococcal enterotoxins. Compendium of Methods for the Microbiological Examination of Foods, 4th edition. Eds American Public Health Association, Downes F.P., Ito K. Washington DC. APHA, 2001, 387-403.

Public Health Laboratory Service. Enumeration of *Staphylococcus aureus*. PHLS standard methods for food products. Public Health Laboratory Service, London. PHLS, 1999, 5.

Bennett R.W., Lancette G.A. *Staphylococcus aureus*. Bacteriological Analytical Manual. Ed. Food and Drug Abministration. 8th edition. Gaithersburg. AOAC International, 1995, 5.

Baird R.M., Lee W.H. Media used in the detection and enumeration of *Staphylococcus aureus*, in *Culture Media for Food Microbiology.* Eds Corry J.E.L., Curtis G.D.W., Baird R.M. Amsterdam. Elsevier, 1995, 77-87.

Steering Group on the Microbiological Safety of Food. Food and environmental methods (10): enumeration of *Staphylococcus aureus*. Ministry of Agriculture, Fisheries and Food. Methods for Use in Microbiological Surveillance. London. MAFF, 1994, 3.

Food and Agriculture Organization, Andrews W. *Staphylococcus aureus*. Manual of Food Quality Control, Vol 4. Microbiological Analysis. Eds Food and Agricultural Organization, Andrews W. Rome. FAO, 1992, 131-8.

VIBRIO

There are some eight species of *Vibrio* that are pathogenic and have been shown to be associated with foods. The main species of *Vibrio* that is of concern to food microbiologists is *Vibrio parahaemolyticus*. However, because of recent concerns, mention will also be made of the organisms *Vibrio cholerae* and *V. vulnificus*.

The Organism *Vibrio parahaemolyticus*

Vibrio parahaemolyticus (originally known as *Pasteurella parahaemolytica*) is a non-sporing, Gram-negative, oxidase and catalase positive, facultatively anaerobic motile rod belonging to the family Vibrionaceae. Cells are often curved or comma-shaped. A particularly significant feature of this organism is the fact that it is a halophile, i.e. it requires a minimum level of salt for growth. It is, therefore, almost exclusively associated with marine environments and seafoods (1,2,3).

Amongst strains of *V. parahaemolyticus*, there appear to be two groups, partly defined by their source and their pathogenic potential. These are the so-called "Kanagawa-positive" and the "Kanagawa-negative" strains. It is the Kanagawa-positive strains that cause *V. parahaemolyticus* food poisoning (2,4,5).

Vibrio parahaemolyticus Food Poisoning

Vibrio parahaemolyticus was first implicated as causing a food-poisoning-type outbreak after the consumption of a fish product in Japan in 1950. Since this time it has become well known as a cause of illness associated with sea fish and other seafoods, especially in the summer months in Japan, where the consumption of raw seafood is a common practice.

Although predominantly associated with Japan, *V. parahaemolyticus* is a potential health hazard associated with seafoods throughout the world.

However, food poisoning from this organism in the UK is a relatively rare phenomenon (the reasons for which will become clear on further reading); where incidents have been reported, they have been largely associated with imported seafoods and foreign travel.

A thermostable hemolysin and toxin are considered major virulence factors of *V. parahaemolyticus*. This haemolysin can be pre-formed in food and is heat-resistant (2,5,6).

Incubation time

The incubation period for *V. parahaemolyticus* food poisoning is between 4 and 96 h (most frequently between 12 and 24 h).

Symptoms

Food poisoning caused by *V. parahaemolyticus* is usually a mild form of gastroenteritis. However, it can begin with violent pains, and diarrhoea occurs in 98% of cases. Nausea also commonly occurs, and other less frequent symptoms include vomiting, headache, mild fever and chills. In severe cases mucus and blood can occur in the stools.

Mortality

The illness is usually self-limiting, lasting about 3-5 days (range 1-7 days), with death being rare.

Infective dose

Volunteer studies indicate that large numbers ($2x10^5$ to $3x10^7$) of Kanagawa-positive *V. parahaemolyticus* cells need to be ingested to induce food-poisoning symptoms (4).

Food involved

As already indicated, foods associated with *V. parahaemolyticus* food poisoning are those of marine origin - fish, shellfish and other seafoods. Gastroenteritis is most commonly associated with the consumption of raw fish, which is a common practice in Japan, or with under-cooked seafoods. Foods cross-contaminated by raw seafoods can also be associated with incidents of gastroenteritis caused by this organism (1,2,3,6).

Because studies indicate that Kanagawa-positive *V. parahaemolyticus* generally occur only in low numbers in foods, it has been concluded that mishandling at temperatures that allow growth of the organism is normally a prerequisite of gastroenteritis caused by this organism (4,6,7,8).

Incidence of *V. parahaemolyticus* Food Poisoning

In the UK, food poisoning caused by *V. parahaemolyticus* is rare, infections in the main being acquired abroad. It was first reported in England and Wales in 1972, but only between 9 and 19 laboratory-reported isolates were recorded each year by CDSC between 1980 and 1989 (9). During 1989-91 there were 69 sporadic cases and over 80% indicated recent foreign travel, with four cases reporting having eaten shellfish while abroad. No cases were reported in Scotland during the 1980s (10).

Other *Vibrio* species (principally *V. cholerae*, other than *V. cholerae* O1) were reported to be the cause of 156 sporadic cases of infection in England and Wales between 1989 and 1991. A similar proportion of these also involved foreign travel. One case of *V. fluvialis* infection was associated with shellfish (10).

As already mentioned, the incidence of foodborne illness associated with this organism is much greater in countries where the consumption of raw fish is common, especially when water temperatures encourage the incidence of vibrios in waters where seafoods are harvested. Thus, in the summer months, the incidence of *V. parahaemolyticus* food poisoning increases in these countries.

In Japan, *Vibrio parahaemolyticus* accounts for over 30% of all annual reported cases of food poisoning. Between 1973 and 1998, a total of 40 outbreaks of *V. parahaemolyticus* infection in the USA were reported to the Centres for Disease Control and Prevention, with >1,000 persons involved (8). They also estimate a 126% increase in the incidence of *Vibrio* infections during the period 1996-2002 (2). This has been partly attributed to the arrival of a new pandemic strain of *V. parahaemolyticus* (O3:K6) during 1996-1998. A *V. parahaemolyticus* (O3:K6) outbreak in 1998, linked to the consumption of raw oysters, involved 416 persons in 13 US states (11).

Sources

Humans

V. parahaemolyticus can be isolated from the stools of asymptomatic individuals in the summer months in Japan, as a transient part of the faecal flora. It can also be excreted for one or two weeks by individuals recovering from *Vibrio*-associated gastroenteritis.

Animals and the environment

V. parahaemolyticus is a halophile and has been isolated from the marine environment throughout the world. Occasional isolations from freshwater or non-marine fish have been reported, but it is thought that this relates to elevated levels of salt, possibly through pollution (7).

Both the incidence and levels of *V. parahaemolyticus* are higher in the summer months; the organism is infrequently found in waters having temperatures less than 10-13 °C. However, the organism may survive the winter months by attaching to chitinous material from plankton (1,3).

It is important to note, however, that most strains of *V. parahaemolyticus* found in the environment are Kanagawa-negative (1).

Foods

Marine shellfish and fish constitute the most important reservoir of *V. parahaemolyticus*. Freshly caught samples may harbour only low numbers of the organism, although in Japan, in the summer, counts of 10^3-10^4/g have been reported. However, most isolates are Kanagawa-negative (7).

Growth/Survival Characteristics of the Organisms in Foods

The most notable characteristic relating to the survival of *V. parahaemolyticus* is its requirement for salt (NaCl) for growth. It is also capable of remarkably rapid growth where conditions are favourable; a generation time as short as 9-13 min has been recorded under optimal conditions (12).

Temperature

The minimum and maximum temperatures reported for growth are 5 °C and 44 °C, with an optimum of 30-35 °C. It fails to grow at 4 °C (1,3,12).

The organism is easily destroyed by drying, and is sensitive to refrigeration temperatures, declining in numbers during storage (7), but only moderately sensitive to freezing (3).

Heat resistance

V. parahaemolyticus is not a heat-resistant organism and is readily destroyed by cooking at temperatures of >65 °C (3,6).D-values at 47 °C range from 0.8-65.1 min has been reported. (1)

pH

This organism is tolerant of high pH values, but less so of low pH values; its pH range for growth is 4.8-11, with an optimum of pH 7.6-8.6 (1)

Water activity/salt

The optimum concentration of NaCl equates to a_w of approximately 0.98. *V. parahaemolyticus* will grow at a_w level as low as 0.922, depending on solute (1).

V. parahaemolyticus is unable to grow in the absence of salt; its upper and lower limits for growth are 8% (although some reports suggest 10) and 0.5% NaCl, respectively, but the optimum concentration is 2-4% (1,3).

Atmosphere

V. parahaemolyticus can grow in either vacuum- or aerobically packed foods (7).

Irradiation

Doses of 3kGy of gamma irradiation can eliminate vibrios from frozen shrimps (4).

Summary of Control of *V. parahaemolyticus* in Foods

Seafoods harvested from waters of sufficiently high temperatures (>10 °C) must be considered to be contaminated with *V. parahaemolyticus*. This organism is able to grow extremely rapidly under conditions of temperature abuse; control, therefore, relies on the strict maintenance of low temperatures.

Adequate cooking of seafood to an internal temperature of >65 °C will destroy any vibrios present.

The prevention of cross-contamination from raw seafoods to prepared foods that are likely to be exposed to temperature abuse is also of paramount importance.

The Organism *Vibrio cholerae*

Like *V. parahaemolyticus*, *V. cholerae* belongs to the family Vibrionaceae; it is a Gram-negative short, straight or curved motile rod, capable of anaerobic growth.

There are a number of serogroups of *V. cholerae*, but until recently it was thought that *V. cholerae* serogroup O1 was the only cause of the disease cholera. However, in mid-1993, it became apparent that a further serogroup was capable of causing cholera; serogroup O139 (4,13). This serogroup has caused large outbreaks in India and Bangladesh, with unusually severe symptoms and is responsible for the eighth pandemic of cholera (1,14).

Vibrio cholerae Illness

Vibrio cholerae causes the illness cholera, which has led to epidemics and pandemics and can be a significant cause of illness and mortality in areas with poor sanitation and hygiene. It is endemic in the subcontinent of India, and in other parts of Asia as well as in Africa. The disease remains a major public health problem in these countries, whereas in western countries the incidence remains low. The recent epidemic in South America is the first epidemic in South America this century (13,15).

Incubation period can range from several hours to five days. The ingestion of sufficient numbers of *V. cholerae* can lead to multiplication of the organism in the intestine and production of cholera toxin (CT) that is an enterotoxin. This results in profuse, watery diarrhoea, involving the excretion of large numbers of organisms. Initially, the stool is brown with faecal matter, but thereafter becomes pale gray with a fishy odour. Mucus in the stool imparts the characteristic 'rice water' appearance. Vomiting is often present. Water loss can be extreme, and resulting dehydration and salt imbalance can, in severe cases, lead to death. Death may occur in as high as 50-70% of the cases (2). However, other illnesses can be produced by strains that lack the ability to produce cholera toxin. In these cases, mild diarrhoea is common (4,13,16).

The infective dose is thought to be usually 10^6-10^8, but can be lower under certain conditions (13,14,17).

Incidence of *V. cholerae* Food Poisoning

In Europe, between January and October 1994, there were 2,339 reported cases of cholera and 47 deaths. This represented a 30-fold increase compared with the previous year (13).

The Pan Health Organisation estimates that during 1991-92 there were 750,000 cases of cholera in South and Central America with 6,500 deaths. More recently, there was a marked increase in cholera cases associated with *V. cholerae* O139 in Bangladesh between March and May 2002, when an estimated 35,000 cases occurred (2).

Sources

Humans can be short-term carriers of *V. cholerae*. Studies have also revealed that household animals like cows, dogs and cats can be sporadic carriers of *V. cholerae* O1 (4).

V. cholerae is normally associated with poor hygiene and faecal contamination; person-to-person or waterborne transmission occurs particularly where there is poor sanitation and/or untreated water supplies. However, food can occasionally become contaminated (4,18).

Fruit and vegetables and other foods can become contaminated by polluted water, the use of 'night soil' as a fertiliser, or through food handlers, the use of contaminated containers, or contamination from flies.

Like *V. parahaemolyticus*, *V. cholerae* is part of the natural marine flora in certain areas, especially estuarine environments. Thus, fish or shellfish may be contaminated if they are harvested from contaminated waters. In these instances, the consumption of raw shellfish can be hazardous.

It is difficult to separate contaminated water from food as a real source of cholera in outbreaks of disease, but several foodborne outbreaks have been reported. These have involved mainly crabs, shrimp, raw fish, mussels, cockles, squid, oysters, clams, rice, raw pork, millet gruel, cooked rice (after three hours without reheating), cooked meat, frozen milk and dairy products, raw fruits and vegetables, and iced candies (4,13,15,18). Other types of foods may also be contaminated with *V. cholerae,* for example, food bought from street vendors, raw seaweed, frozen fresh coconut milk, palm fruit, cooked potatoes, eggs, pasta and spices (2).

Growth/Survival Characteristics of the Organism in Foods

Unlike *V. parahaemolyticus*, *V. cholerae* may not have an absolute requirement for NaCl, although its growth is enhanced by the presence of low concentrations of salt. It will grow in up to 4% NaCl (16).

Its optimum temperature for growth has been reported as 30-37 °C, with a temperature range for growth of 10-43 °C. Like *V. parahaemolyticus*, it is tolerant of high pH values, but not acid conditions; it has a pH range for growth of pH 5-9.6 with an optimum value of 7.6 (13,21).

The organism can grow in some foods, including seafoods, certain cooked foods (where there are minimal competitors) and certain raw vegetables. *V. cholerae* can survive in moist, low-acid chilled foods for 2 or more weeks (it has been reported to survive on shellfish for 2-6 weeks, depending on temperature, and will survive for long periods on frozen shellfish); survival in high-acid foods is usually less than 1 day, and in dry foods less than 2 days (15).

The organism is sensitive to the usual methods of disinfection used in food processing and preparation, including heat, drying and chlorination. It is not a heat-resistant organism, being killed by pasteurisation temperatures, and cooking to 70 °C is normally adequate to ensure the destruction of *V. cholerae* (18).

Drying and exposure to sunlight is also an effective means of killing *V. cholerae*. *V. cholerae* can survive domestic freezing and can be found after a long period in a frozen state (2). Treatment of fresh and frozen frog legs with >50 krad of ^{60}Co is known to eliminate *V. cholerae* (4).

Note: Not all strains of serogroup O1 and O139 produce CT, cause cholera or are involved in human illness. Therefore, in assessing public health significance, two critical properties must be determined. Firstly, the production of CT and secondly, the possession of the O1 and O139 antigen (4).

The Organism *Vibrio vulnificus*

This organism does not currently appear to be causing any significant concern in the UK, although some cases have been reported in Europe. It has, however, received considerable attention in the USA because of the high fatality rate (95%). *V. vulnificus* infections have been associated with certain raw seafoods, especially oysters (8,12,19).

V. vulnificus Food Poisoning

Vibrio vulnificus is highly invasive, and can cause septicaemia, which has a high associated mortality rate (approximately 60%), amongst susceptible people, through the ingestion of contaminated raw seafood (8). Wound infections can occur if seafood handlers cut themselves while cleaning or harvesting oysters in such a way that the wound becomes contaminated with sea water (4).

The incubation period for septicaemia after ingestion of the organism is generally between 16 and 38 h, and symptoms most commonly involve fever and chills. Nausea, abdominal pain, vomiting, diarrhoea and hypotension occur less frequently. Progression of the illness can be very rapid, from asymptomatic to death within 24 h. Secondary lesions often occur, especially in the extremities, which may require major surgical attention (4,8,19).

Individuals who suffer septicaemia after the ingestion of *V. vulnificus* almost always have some underlying chronic disease, especially a liver- or blood-related disorder, which may be linked with iron metabolism. Most infections occur in males over the age of 50, during the summer months (1,4,19).

Illness caused by *V. vulnificus* may be associated with the production of a toxin or haemolysin, but other virulence factors, including a protease and other enzymes, as well as haemagglutinating activity and a polysaccharide capsule may also be inloved. The role of the capsule in the pathogenesis of human infection appears to be clearly established (4,12,20).

Sources

Principal reservoirs for *V. vulnificus* are coastal seawater and brackish water (12).

V. vulnificus has been detected in clams and oysters taken from the Gulf, East and Pacific coasts of the USA, and from around the world. Raw oysters constitute a major food vehicle for the transmission of *V. vulnificus* infection in the USA (20,21).

Growth/Survival Characteristics of the Organism in Foods

V. vulnificus is a halophile, having an absolute requirement for salt. Its optimum NaCl concentration for growth is 2.5%, and it is able to grow at concentrations of 0.5-5.0% NaCl (20) but not at 0% or 8%.

The lack of association between the consumption of cooked oysters and illness suggests that *V. vulnificus* is heat-sensitive. In fact, research has shown that cooking oysters for 10 min at 50 °C should ensure the destruction of the organism (22). D-values at 47 °C for 52 strains of *V. vulnificus* average 78 sec (4).

It is also susceptible to ionisation and high hydrostatic pressure. A dose of 1kGy is known to decrease *V. vulnificus* populations in shell stock oysters by greater than 5 log cfu/g. Higher doses of 1.5 kGy completely inactivate *V. vulnificus* (4).

The optimum growth temperature for this organism is 37 °C, although growth may occur within the range 8-43 °C. *V. vulnificus* is sensitive to low temperatures; its presence has been observed mainly during warm months, rarely in cold waters or in chilled seafood, such as iced oysters. However, research suggests that this organism enters a 'viable but non-culturable state' in the cold environment, so that failure to isolate *V. vulnificus* from cold samples may not be a reliable measure of its absence (8,12).

The organism may be a part of the normal flora of certain estuarine environments, and its presence does not appear to correlate with the presence of indicator bacteria (21). Its numbers are related to water temperature, and a 1994 survey found it to be common in Danish marine environments during an unusually warm summer, indicating that it is probably ubiquitous (23).

Bibliography

References

1. Jay J.M., Loessner M.J., Golden D.A. Foodborne gastroenteritis caused by *Vibrio*, *Yersinia*, and *Campylobacter* species, in *Modern Food Microbiology*. Eds Jay J.M., Loessner M.J., Golden D.A. New York. Springer Science, 2005, 657-78.

2. Nair G.B, Faruque S.M, Sack D.A. Vibrios, in *Emerging Foodborne Pathogens*. Eds Motarjemi Y, Adams M. Cambridge. Woodhead Publishing Ltd., 2006, 332-72.

3. International Commission on Microbiological Specifications for Foods. *Vibrio parahaemolyticus*, in Microorganisms in Foods, Vol. 5. Microbiological Specifications of Food Pathogens. International Commission on Microbiological Specifications for Foods. London. Blackie, 1996, 426-35.

4. Oliver J.D., Kaper J.B. *Vibrio* Species, in *Food Microbiology: Fundamentals and Frontiers*. Eds Doyle M, Beuchat L, Montville T. Washington DC. ASM Press, 2001, 263-300.

5. Nishibuchi M., DePaola A. *Vibrio* species, in *Foodborne pathogens: microbiology and molecular biology*. Eds Fratamico P.M., Bhunia A.K., Smith J.L. Wymondham. Caister Academic Press, 2005, 251-71.

6. Desmarchelier P.M. Pathogenic vibrios, in Foodborne Microorganisms of Public Health Significance. Ed. Australian Institute of Food Science and Technology. 5th edition. North Sydney. AIFST, 1997, 285-312.

7. Twedt R.M. *Vibrio parahaemolyticus*, in *Foodborne Pathogens*. Ed. Doyle M.P. New York. Marcel Dekker, 1989, 543-68.

8. Montville T.J, Matthews K.R. *Vibrio* species, in *Food microbiology: an introduction*. Eds Montville T.J., Matthews K.R. Washington DC. ASM Press, 2005, 147-56.

9. Committee chaired by Sir Mark Richmond. The Microbiological Safety of Food Report, Part II. London. HMSO, 1991.

10. Sockett P.N., Cowden J.M., Le Baigue S., Ross D., Adak G.K., Evans H. Foodborne disease surveillance in England and Wales: 1989-1991. *CDR Review,* 1993, 3 (12), R159-73.

11. Daniels N.A. *et al.* Emergence of a new *Vibrio parahaemolyticus* serotype in raw oysters. *Journal of the American Medical Association,* 2000, 284 (12), 1541-5.

12. Sakazaki R., Kaysner C., Abeyta C. *Vibrio* infections, in *Foodborne infections and intoxications*. Eds Riemann H.P, Cliver D.O. London. Academic Press, 2005, 185-204.

13. Crowcroft N.S. Cholera: current epidemiology. *CDR Review,* 1994, 4 (13), R157-64.

14. Reen F., Boyd E. *Vibrio* species: pathogenesis and stress response, in *Understanding pathogen behaviour: virulence, stress response and resistance.* Ed. Griffiths M. Cambridge. Woodhead Publishing Ltd., 2005, 358-88.

15. Popovic T., Olsvik O., Blake P.A., Wachsmuth K. Cholera in the Americas: Foodborne aspects. *Journal of Food Protection,* 1993, 56 (9), 811-21.

16. International Commission on Microbiological Specifications for Foods. *Vibrio cholerae*,in Microorganisms in Foods, Vol. 5. Microbiological Specifications of Food Pathogens. International Commission on Microbiological Specifications for Foods. London. Blackie, 1996, 414-25

17. Management of Outbreaks of Foodborne Illness. *Vibrio cholerae* serotype O1 and non O1. Ed. Department of Health.. London. HMSO. 1994, 91-2.

18. Roberts D. Growth and survival of *Vibrio cholerae* in foods. *PHLS Microbiology Digest*, 1992, 9 (1), 24-31.

19. Sutherland J., Varnam A. Enterotoxin producing *Staphylococcus*, *Shigella*, *Yersinia*, *Vibrio*, *Aeromonas* and *Plesiomonas*, in *Foodborne pathogens: Hazards, risk analysis and control.* Cambridge. Woodhead Publishing Limited, 2002, 401-7.

20. International Commission on Microbiological Specifications for Foods. *Vibrio vulnificus*, in Microorganisms in Foods, Vol. 5. Microbiological Specifications of Food Pathogens. International Commission on Microbiological Specifications for Foods. London. Blackie, 1996, 436-9.

21. Oliver J.D., Warner R.A., Cleland D.R. Distribution of *Vibrio vulnificus* and other lactose-fermenting vibrios in the marine environment. *Applied and Environmental Microbiology,* 1983, 45 (3), 985-98.

22. Cook D.W., Ruple A.D. Cold storage and mild heat treatment as processing aids to reduce the numbers of *Vibrio vulnificus* in raw oysters. *Journal of Food Protection,* 1992, 55 (12), 985-9.

23. Hoi L., Larsen J.L., Dalsgaard I., Dalsgaard A. Occurrence of *Vibrio vulnificus* biotypes in Danish marine environments. *Applied and Environmental Microbiology,* 1998, (January), 64 (1), 7-13.

Further reading

Swaminathan B., Gerner-Smidt P., Whichard J.M. Foodborne disease trends and reports, in *Foodborne Pathogens and Disease,* 2006, (Winter), 3 (4), 316-8.

Food and Agriculture Organization, World Health Organization. Risk assessment of *Vibrio vulnificus* in raw oysters: interpretative summary and technical report. *Microbiological Risk Assessment Series*. Rome. FAO, 2006.

Vongxay K., He X., Cheng S., Zhou X., Shen B., Zhang G., Zhang J., Fang W. Prevalence of *Vibrio parahaemolyticus* in seafoods and their processing environments as detected by duplex PCR. *Journal of the Science of Food and Agriculture,* 2006, (September), 86 (12), 1871-7.

Bang W., Drake M.A. Acid adaptation of *Vibrio vulnificus* and subsequent impact on stress tolerance. *Food Microbiology,* 2005, (August), 22 (4), 301-9.

Miliotis M., Bier J. International Handbook of Foodborne Pathogens. New York. Marcel Dekker Inc., 2003.

Oliver J.D., Kaper J.B. *Vibrio* species, in *Food Microbiology: Fundamentals and Frontiers.* Eds Doyle M.P., Beuchat L.R., Montville T.J. 2nd edition. Washington DC. ASM Press, 2001, 263-300.

Tamplin M.L. *Vibrio vulnificus*, *Vibrio parahaemolyticus*, and *Vibrio cholerae*, in *Guide to Foodborne Pathogens.* Eds Labbe R.G., Garcia S. New York. Wiley, 2001, 221-43.

Kaysner C.A. *Vibrio* species, in *The Microbiological Safety and Quality of Food, Volume 2.* Eds Lund B.M., Baird-Parker T.C., Gould G.W. Gaithersburg. Aspen Publishers, 2000, 1336-62.

Kaysner C.A., Wetherington J.H., Chai T.-J., Pace J.L., Dalsgaard A., Hoi L., Linkous D., Oliver J.D. *Vibrio cholerae, Vibrio parahaemolyticus, Vibrio vulnificus*, in *Foodborne disease handbook, volume 1: bacterial pathogens.* Eds Hui Y.H., Pierson M.D., Gorham J.R. 2nd edition. New York. Marcel Dekker, 2000, 383-470.

Desmarchelier P.M. Pathogenic vibrios. Foodborne Microorganisms of Public Health Significance. Ed. Australian Institute of Food Science and Technology. 5th edition. North Sydney. AIFST, 1997, 285-312.

International Commission on Microbiological Specifications for Foods. *Vibrio cholerae*, *Vibrio parahaemolyticus*, *Vibrio vulnificus*, in Microorganisms in Foods, Vol. 5. Microbiological Specifications of Food Pathogens. International Commission on Microbiological Specifications for Foods. London. Blackie, 1996, 414-39.

Methods of detection

Fyske E.M., Skogan G., Davies W., Olsen J.S., Blatny J.M. Detection of *Vibrio cholerae* by real-time nucleic acid sequence-based amplification. *Applied and Environmental Microbiology,* 2007, (March), 73 (5), 1457-66.

Nordstrom J.L., Rangdale R., Vickery M.C.L., Phillips A.M.B., Murray S.L., Wagley S., DePaola A. Evaluation of an alkaline phosphatase-labeled oligonucleotide probe for the detection and enumeration of the thermostable-related hemolysin (trh) gene of *Vibrio parahaemolyticus*. *Journal of Food Protection,* 2006, (November), 69 (11), 2770-2.

Di Pinto A, Ciccarese G, Fontanarosa M., Terio V, Tantillo G. Detection of *Vibrio alginolyticus* and *Vibrio parahaemolyticus* in shellfish samples using collagenase-targeted multiplex-PCR. *Journal of Food Safety,* 2006, (May), 2 (2), 150-9.

Ward L.N., Bej S.K. Detection of *Vibrio parahaemolyticus* in shellfish by use of multiplexed real-time PCR with TaqMan fluorescent probes. *Applied and Environmental Microbiology,* 2006, (March), 72 (3), 2031-42.

Wang S., Levin R.E. Rapid quantification of *Vibrio vulnificus* in clams (Protochaca staminea) using real-time PCR. *Food Microbiology,* 2006, (December), 23 (8), 757-61.

Sanath Kumar H., Parvathi A., Karunasagar I., Karunasagar I. A gyrB-based PCR for the detection of *Vibrio vulnificus* and its application for direct detection of this pathogen in oyster enrichment broths. *International Journal of Food Microbiology,* 2006, (October 1), 111 (3), 216-20.

Su Y.-C., Duan J., Wu W.-H. Selectivity and specificity of a chromogenic medium for detecting *Vibrio parahaemolyticus*. *Journal of Food Protection,* 2005, (July), 68 (7), 1454-6.

Wang S., Levin R.E. Quantification of *Vibrio vulnificus* using the polymerase chain reaction. *Food Biotechnology,* 2005, 19 (1), 27-35.

Kaysner C.A., DePaola A. *Vibrio*, in Compendium of Methods for the Microbiological Examination of Foods. American Public Health Association. Eds Downes F.P., Ito K. 4th edition. Washington DC. APHA, 2001, 405-20.

Andrews W.H. Microbiological methods. Subchapter 11. *Vibrio*, in Official Methods of Analysis of AOAC International, Volume 1: Agricultural Chemicals, Contaminants, Drugs. AOAC International, Horwitz W. 17th edition. Gaithersburg. AOAC International, 2000, Chapter 17, 164-8.

Arias C.R., Aznar R., Pujalte M.J., Garay E. A comparison of strategies for the detection and recovery of *Vibrio vulnificus* from marine samples of the western Mediterranean coast. *Systematic and Applied Microbiology,* 1998 (March), 21 (1), 128-34.

Koch W.H., Payne W.L., Cebula T.A. Detection of enterotoxigenic *Vibrio cholerae* in foods by the polymerase chain reaction, in Bacteriological Analytical Manual. Ed. Food and Drug Administration. 8th edition. Gaithersburg. AOAC International, 1995, 9.

Elliott E.L., Kaysner C.A., Jackson L., Tamplin M.L. *V. cholerae, V. parahaemolyticus, V. vulnificus* and other *Vibrio* spp., in Bacteriological Analytical Manual. Ed. Food and Drug Administration. 8th edition. Gaithersburg. AOAC International, 1995, 27.

Donovan T.J., van Netten P. Culture media for the isolation and enumeration of pathogenic *Vibrio* species in foods and environmental samples.*International Journal of Food Microbiology,* 1995, 26 (1), 77-91.

Hagen C.J., Sloan E.M., Lancette G.A., Peeler J.T., Sofos J.N. Enumeration of *Vibrio parahaemolyticus* and *Vibrio vulnificus* in various seafoods with two enrichment broths. *Journal of Food Protection,* 1994, 57 (5), 403-9.

Food and Agriculture Organization, Andrews W. *Vibrio cholerae*, *Vibrio parahaemolyticus*, in Manual of Food Quality Control, Vol. 4. Microbiological Analysis. Eds Food and Agriculture Organization, Andrews W. Rome. FAO, 1992, 57-78.

PATHOGENIC *ESCHERICHIA COLI* (VTEC)

VTEC (vero cytotoxigenic *E. coli*) was first implicated in infectious disease in 1982, when it was associated with two outbreaks of haemorrhagic colitis (Riley *et al.* 1983) involving hamburger patties in sandwiches. Since then, it has caused and continues to cause considerable concern in developed countries; there have been several serious foodborne outbreaks in the UK and in other other developed countries.

The Organism VTEC

VTEC in common with other *E. coli* are facultative anaerobic, Gram negative non-spore-forming, catalase-positive, oxidase-negative, motile rods. It can be broken down into a number of serotypes, which produce their own particular toxin.

The term 'vero cytotoxigenic *E. coli*' (VTEC) encompasses those strains of *E. coli* that are capable of producing one or more toxins that are cytotoxic to Vero cells - a tissue cell culture line established from the kidneys of an African Green monkey. These toxins are known as vero cytotoxins (or verotoxins). They are also sometimes referred to as 'Shiga-like' toxins, particularly in North America (2).

More than 200 VTEC serotypes have been isolated from humans (source: WHO), but in many countries the main serogroup that has been demonstrated to produce symptoms associated with illness caused by VTEC is *E. coli* O157, termed O157 VTEC. The main serotype of concern in this group is O157:H7. O157 VTEC can also cause haemorrhagic colitis and can therefore be described as EHEC. Not all VTEC serotypes are EHEC, but other important non-O157 EHEC serogroups associated with foodborne illness include O26, O111, O103 and O145 (3,4).

Recently, antibiotic resistant strains of VTEC have been isolated from foods. Investigation into the relationship between antibiotic resistance and death and growth kinetics in various foods, showed no difference in growth kinetics, but in acid foods (yoghurt and orange juice) the multiple antibiotic resistant strains died off faster (5).

Verotoxigenic *E. coli* Food Poisoning

The organism *E. coli* O157:H7 was first associated with foodborne illness in 1982, in the USA. In this incident, it was discovered in stool samples from patients suffering haemorrhagic colitis, who had eaten hamburgers at a fast-food restaurant. It was also isolated from a hamburger sample taken from the implicated lot (6).

Since 1982, VTEC has been increasingly linked with incidents of haemorrhagic colitis (HC - a severe illness involving bloody diarrhoea) and haemolytic uraemic syndrome (HUS - a kidney disease). VTEC infection involving progression to HUS is most common in infants and young children.

Incubation time

The incubation period is unusually variable, and may range from 1-14 days, but is typically between 3-4 days.

Symptoms

Illness caused by VTEC can take the form of relatively mild diarrhoea only. However, it can cause serious illness, particularly in vulnerable groups. Typical illness begins with non-bloody diarrhoea and severe abdominal cramps. Vomiting can occur. In about half of cases, diarrhoea becomes bloody (HC) by about two days after onset of illness; this may be more blood than stool and may be mistaken for intestinal bleeding. A small proportion of VTEC cases (between 2 and 7%, but up to 30%) may go on to develop HUS. A further condition that may result from infection by VTEC is thrombotic thrombocytopaenic purpura (TTP), an extension of HUS with additional fever and neurological symptoms. TTP mainly affects adults rather than children and can be very serious.

Mortality

VTEC is the leading cause of acute renal failure in children, which can be fatal. Also, fatalities have been reported in elderly patients with underlying medical problems. Those at greatest risk from VTEC are those between 2 and 10 years, and those over 60 years of age. The average mortality rate as a result of VTEC-associated HUS infection is reported as 1% (7,8).

Infective dose

The infective dose for VTEC is thought to be very low, and has been reported as between 2-2,000 cells for *E. coli* O157:H7 (9). Intact packages of salami associated with a foodborne outbreak were found to contain 0.3-0.4 *E. coli* O157 cells per g (10). In another outbreak associated with salami, samples of the incriminated product were found to contain *E. coli* O111 at a level of less than one cell per 10 g (11).

Foods involved

In recent years, foodborne outbreaks of VTEC infection, particularly by serogroup O157, have become a serious public health problem. Although foodborne outbreaks have occurred in a number of countries, the incidence of infection seems to be greater in North America, the UK and mainland Europe. Large outbreaks have also been recorded in Japan and South Africa, and Latin America (particularly Argentina) reportedly has a high incidence. VTEC strains of serogroup O157 were first isolated in the UK in 1983 following an HUS outbreak. Since then, there have been at least 150 general outbreaks in England and Wales up to the end of 2001 (source: PHLS). Epidemiological evidence suggests that many of these resulted from the consumption of contaminated foods, including beef, cooked meat, raw and pasteurised (recontaminated) milk, cheese and raw vegetables. Laboratory reports of O157 VTEC infection rose steadily after 1983, peaking at 1,087 cases in 1997 (source: PHLS) (9). Similar increases have been seen in North America. The most serious outbreak to date occurred in central Scotland in late 1996. The outbreak resulted in 512 cases and 22 deaths, and was associated with cooked meats supplied by a local butcher (12).

In North America, VTEC infection has been linked mainly with undercooked ground beef and, to a lesser extent, raw milk. Thus, in the USA and Canada, VTEC is generally considered to be of (dairy) bovine origin. In a major outbreak in January 1993, undercooked, contaminated beefburgers led to an outbreak involving 732 reported cases and four deaths, centred on Washington State (13). However, the most recent outbreak involving VTEC in the USA was in 2006, and was associated with spinach. This resulted in 205 confirmed illnesses and three deaths.

In Japan, a major outbreak (the largest recorded) occurred in 1996, in which over 9,000 people may have been infected, with 10 subsequent deaths. Most of those involved were children and the outbreak was associated with radish sprouts used as a component of school lunches (14).

Other foods that have been associated with outbreaks in several countries include yoghurt, mayonnaise, unfermented apple cider, unpasteurised apple juice, cantaloupe, salami and alfalfa sprouts (15).

Incidence of VTEC Food Poisoning

The incidence of VTEC infection in the UK shows a geographical variation. Between 1990 and 1996, the average annual rate in England and Wales was approximately 1.5/100,000, but in Scotland the average rate was over 5/100,000 for the same period. The incidence in England and Wales increased steadily until reaching a peak of 1,087 cases in 1997 (source: PHLS). Since then, the number of reported cases remained at approximately 900-1,000 per year until 2002, when only 595 cases were recorded. The UK generally has a higher reported incidence of O157 VTEC infection than other EU countries.

High rates have also been recorded in the USA and Canada. The US incidence of O157 VTEC infection in 2001 was 1.6 cases per 100,000 people, but this represented a decrease of 21% compared with the previous four years (16). In Canada, an isolation rate of 5.2/100,000 was reported in 1987; this increased to 8.8/100,000 in 1989, but has since decreased and the incidence in 1995 was reported as 5.1/100,000 (17).

Sources

Humans

There is evidence of person-to-person spread in such places as institutions and child-care centres, and some secondary transmission from person-to-person may occur during foodborne outbreaks. Some individuals may also act as asymptomatic carriers (3). It is reported that people who are professionally exposed to cattle can develop resistance to VTEC infection (18).

Animals and environment

There is strong evidence that cattle are a major reservoir for O157 VTEC. In the UK, a study reported in 1997 showed that 15.7% of rectal swabs from 4,800 cattle were positive for the organism. It was also found in 2.2% of sheep, but not in pigs or poultry (19), although other studies have isolated the organism from these species. A comparison of strains from cattle and local human cases in the Sheffield area showed a strong correlation (20). Faecal shedding of VTEC by animals appears to be seasonal. A recent study showed that *E. coli* O157:H7 could be isolated from the faeces of 38% of cattle in the spring, but this figure fell to 4.8% in winter (11).

A wide range of other animals have also been found to excrete VTEC, including goats, horses, dogs, cats, deer, rats and seagulls. *E. coli* O157:H7 has even been isolated from houseflies (21). There have been cases of infection and outbreaks associated with direct contact with animals, often linked to petting zoos.

It has been shown that bovine hides are a significant vector for the transmission of VTEC, onto prepared carcases within abattoirs (22).

There have also been a number of large outbreaks associated with water supplies, particularly wells contaminated with animal faeces. In 2000, a major outbreak in Canada resulted in over 1,000 cases and a number of deaths, and was linked to a contaminated mains water supply (23).

Foods

North American studies have revealed VTEC - including, occasionally, O157 - in a small proportion of various raw meat samples, and indicated a low incidence in raw milk. The food from which *E. coli* O157:H7 has been most frequently isolated is raw ground beef, especially dairy beef. It has been noted particularly in Canada, where the incidence of *E. coli* O157:H7 infection is notably high. Results from the US Department of Agriculture (USDA) microbiological testing programme for raw ground beef products show an increase in the number of positive samples from 0.06% in 1995 to 0.8% in 2001. Of the 171 positive samples recorded from 1994-2001, 114 were recorded during the period from 2000-2001 (source: USDA Food Safety and Inspection Service).

In the UK, *E. coli* O157 appears to be rare in foods. A survey reported in 1996 showed that O157 VTEC was found in 3.4% of frozen beefburgers sampled and in 2.0% of fresh minced beef, but it was not found in fresh sausages (24). A second survey, carried out in 1996-7, found that O157 VTEC could be isolated from 1.1% of raw beef products and 2.9% of raw lamb products, especially lamb sausages and lamb burgers, although it was suspected that these may have been contaminated by other added ingredients (25).

Survey results of animal carcases and meats to date generally suggest that the organism in the UK is principally of bovine origin. Thus undercooked beef and raw milk are likely to be the prime 'high-risk' foods in this country. However, the variety of foods implicated in outbreaks indicates that many products may potentially become contaminated. A summer peak in infections also suggests a possible link with barbecued (undercooked) meat products and/or a higher rate of infection in cattle associated with pasture feeding.

Growth/Survival Characteristics of the Organism in Foods

Temperature

Under otherwise favourable conditions, the lowest reported temperature for growth of *E. coli* O157, in laboratory media, is approximately 7 °C, and the highest is 45 °C, with an optimum of 37 °C (2). (Note *E. coli* O157:H7 grows poorly at 44-45 °C and does not grow within 48 h at 45.5 °C. Therefore, traditional detection methods for *E. coli* in foods cannot be relied upon to detect *E. coli* O157:H7.). Other non-O157 VTEC strains, particularly O26, are also reported to be inhibited by high temperatures (26).

The organism appears to survive well at low temperatures and to resist freezing.

Heat resistance

E. coli O157 is not a heat-resistant organism. $D_{57\,°C}$ and $D_{63\,°C}$ values of approximately 5 min and 0.5 min, respectively, in meat, have been reported (27). Anaerobic growth, reduced a_w, high fat content and exposure to prior heat shock may result in higher D-values. The cooking of beefburgers to

70 °C for 2 min or equivalent in all parts of every beefburger has been recommended in the UK (28).

pH

The minimum pH for growth of *E. coli* O157:H7, under otherwise optimal conditions, is reported to be 4.0-4.4 (29). The minimum value is affected by the acidulant used, with both acetic and lactic acids being more inhibitory than hydrochloric acid (2). *E. coli* O157 is unusually acid-tolerant and survives well at low pH values, as witnessed by its survival in American unfermented 'apple cider' (pH 3.6-4.0) and mayonnaise (reported to be pH 3.8) - two apparent vehicles of infection in two outbreaks in the USA - especially at chill temperatures (15).

Water activity/salt

Current published data suggest that *E. coli* O157 grows well at NaCl concentrations up to 2.5% and may grow at NaCl concentrations of at least 6.5% (w/V) (a_w less than 0.97) under otherwise optimal conditions (30). A recent outbreak associated with dry fermented meat (a_w approximately 0.9) demonstrated the ability of the organism to survive the fermentation and drying processes used in the manufacture of such products (15). *E. coli* O157:H7 is highly resistant to desiccation (31).

Atmosphere

E. coli O157 is a facultative anaerobe; it grows well under aerobic or anaerobic conditions. Modified-atmosphere packaging had little effect on the growth or survival of this organism (32).

Irradiation

VTEC strains do not differ significantly from other Gram-negative pathogens in their resistance to gamma irradiation. Typical D-values of 0.24-0.39 kGy have been recorded in chilled meat and poultry (33).

The use of irradiation to control O157 VTEC in beef products has been proposed in the US. In 2000, this treatment was permitted by the USDA for chilled and frozen red meat, and maximum dose levels were specified. Irradiated ground beef products have recently gone on sale at some retail outlets.

Summary of Control of VTEC in Foods

Control measures against VTEC are similar to those for other vegetative pathogens, such as *Salmonella* and *Listeria*. The fact that it appears primarily to be associated with bovine sources means that greatest attention should be paid to the avoidance of unpasteurised milk and to the hygiene practices in beef processing. Contamination with faecal material must be minimised; cross-contamination from raw to cooked products must be avoided and adequate cooking should be employed (where appropriate) to ensure the destruction of any VTEC present. As in the case of other pathogens, the HACCP approach should be used to assist in these measures. There is still a need for further research into the effect of food processes on the survival and growth of VTEC.

Other Pathogenic *E. coli*

Escherichia coli is part of the normal flora of the intestinal tract of both man and warm-blooded animals. The presence of *E. coli* in water or foods, therefore, may indicate faecal contamination, although their consumption may not lead to any apparent ill effects on health. Certain serotypes of *E. coli*, however, may cause diarrhoeal disease, or more serious forms of illness. These can be divided into several virulence groups, described as enteropathogenic (EPEC), enterotoxigenic (ETEC), enteroinvasive (EIEC), diffusely adherent (DAEC), enteroaggregative (EAggEC) or vero cytotoxigenic (VTEC) (also described as Shiga-like toxin-producing (STEC)). The last group includes those serotypes described as enterohaemorrhagic (EHEC).

In developing countries, diarrhoea amongst infants and young children caused by different types of pathogenic *E. coli* is a major cause of morbidity and mortality. In developed countries, where standards of hygiene are high, the pathogenic *E. coli* are of relatively minor importance; little diarrhoeal illness has been associated with any *E. coli*, other than 'travellers' diarrhoea', in which enterotoxigenic *E. coli* are important (1). Verotoxigenic *E. coli*, particularly serogroup O157, however, is causing considerable current concern in developed countries. The increasing incidence of *E. coli*

O157 infections, including several very serious foodborne incidents in the UK, has increased concerns over the organism in this country. Further details in this Micro-Facts will, therefore, be restricted to this organism - vero cytotoxigenic *E. coli* (VTEC).

Bibliography

References

1. Doyle M.P., Padhye V.V. *Escherichia coli*, in *Foodborne Bacterial Pathogens.* Ed. Doyle M.P. New York. Marcel Dekker, 1989, 235-81.

2. Advisory Committee on the Microbiological Safety of Food. Report on verocytotoxin-producing *Escherichia coli.* London. HMSO, 1995.

3. Bell C. Approach to the control of enterohaemorrhagic *Escherichia coli* (EHEC). *International Journal of Food Microbiology,* 2002, (October 25), 78 (3), 197-216.

4. Many authors. Community outbreak of hemolytic uremic syndrome attributable to *E. coli* O111:NM - South Australia, 1995. *Morbidity and Mortality Weekly Report*, 1995, 44 (29), 550-1+557-8.

5. Duffy G., Walsh C., Blair I.S., McDowell D.A. Survival of antibiotic resistant and antibiotic sensitive strains of *E. coli* O157 and *E. coli* O26 in food matrices. *International Journal of Food Microbiology,* 109 (2006), 179-86.

6. Riley L.W., Remis R.S., Helgerson S.D. *et al.* Haemorrhagic colitis associated with a rare *Escherichia coli* serotype. *New England Journal of Medicine,* 1983, 308 (12), 681-5

7. Anon. The prevention of human transmission of gastrointestinal infections, infestations and bacterial intoxications, *Escherichia coli*-vero cytotoxin producing (VTEC). *Communicable Disease Report*, 1995, 5 (11), R164.

8. Doyle M.P., Zhao T., Meng J., Zhao S. *Esherichia coli* O157:H7, in *Food Microbiology: Fundamentals and Frontiers.* Eds Doyle M.P., Beuchat L.R., Montville T.J. Washington DC. ASM Press, 1997, 171-91.

9. Buchanan R.L., Doyle M.P. Foodborne disease significance of *Escherichia coli* O157:H7 and other enterohemorrhagic *E. coli. Food Technology,* 1997, (October), 51 (10), 69-76

10. Anon. *Escherichia coli* O157:H7 outbreak linked to commercially distributed dry-cured salami - Washington and California, 1994. *Morbidity and Mortality Weekly Report,* 1995, 44 (9), 157-60.

11. Meng J., Doyle M.P., Zhao T., Zhao S. Enterohemorrhagic *Escherichia coli,* in *Food microbiology: fundamentals and frontiers.* Eds Doyle M.P., Beuchat L.R., Montville T.J. 2nd edition. Washington DC. ASM Press, 2001, 193-213.

12. Cowden J.M., Ahmed S., Donaghy M., Riley A. Epidemiological investigation of the Central Scotland outbreak of *Escherichia coli* O157 infection, November to December 1996. *Epidemiology and Infection,* 2001 (June), 126 (3), 335-41.

13. Bell B.P., Goldoft M., Griffin P.M., Davis M.A., Gordon D.C., Tarr P.I., Bartleson C.A., Lewis J.H., Barrett T.J., Wells J.G., Baron R., Kobayashi J. A multistate outbreak of *Escherichia coli* 0157:H7-associated bloody diarrhea and hemolytic uremic syndrome from hamburgers. The Washington experience. *Journal of the American Medical Association (JAMA)*, 1994, 272 (17), 1349-53

14. Izumya H., Terajima J., Wada A. *et al.* Molecular typing of enterohaemorrhagic *Escherichia coli* O157:H7 isolates in Japan by using pulsed-field gel electrophoresis. *Journal of Clinical Microbiology,* 1997, 35, 230-7.

15. Meng J., Doyle M.P. Microbiology of Shiga-toxin-producing *Escherichia coli* in foods, in *Escherichia coli* O157:H7 and Other Shiga Toxin-producing *E. coli* Strains. Eds Kaper J.P., O'Brien A.D. Washington DC. *American Society for Microbiology*, 1998, 92-108.

16. Vugia D., Hadler J., Blake P., Blythe D., Smith K., Morse D., Cieslak P., Jones T., Shillam P., Chen D.W., Garthright B., Charles L., Molbak K., Angulo F., Griffin P., Tauxe R. Preliminary FoodNet data on the incidence of foodborne illnesses - selected sites, United States, 2001. *Morbidity and Mortality Weekly Report (MMWR)*, 2002 (April 12), 51 (14), 325-9.

17. Division of Disease Surveillance. Notifiable Diseases Annual Summary 1995. *Canada Communicable Disease Report*, 1997, 23S9.

18. Silvestro L. *et al.* Asymptomatic carriage of verotoxin-producing *Escherichia coli* O157 in farm workers in Northern Italy. *Epidemiology and Infection,* 2004 (October), 132, 915-9.

19. Chapman P.A., Siddons C.A., Cerdan Malo A.T., Harkin M.A. A 1-year study of *Escherichia coli* O157 in cattle, sheep, pigs and poultry. *Epidemiology and Infection,* 1997 (October), 119 (2), 245-50.

20. Chapman P.A., Wright D.J., Norman P. Verotoxin-producing *Escherichia coli* infections in Sheffield: cattle as a possible source. *Epidemiology and Infection,* 1989, 102 (3), 439-45.

21. Kobayashi M., Sasaki T., Saito N., Tamura K., Suzuki K., Watanabe H., Agui N. Houseflies: not simple mechanical vectors of enterohemorrhagic *Escherichia coli* O157:H7. *American Journal of Tropical Medicine and Hygiene,* 1999, 61 (4), 625-9.

22. O'Brien S.B., Duffy G., Carney E., Sheridan J.J., McDowell D.A., Blair I.S. Prevalence and Numbers of *Escherichia coli* O157 on Bovine Hides ata beef Slaughter Plant. *Journal of Food Protection,* 2005, 68 (4), 660-65.

23. Spurgeon D. Budget cuts may have led to *E. coli* outbreak. *British Medical Journal* (BMJ), 2000 (June 17), 7250 (320), 1625.

24. Bolton F.J., Crozier L., Williamson J.K. Isolation of *Escherichia coli* O157 from raw meat products. *Letters in Applied Microbiology,* 1996, 23 (5), 317-21.

25. Chapman P.A., Siddons C.A., Cerdan Malo A.T., Harkin M.A. A one year study of *Escherichia coli* O157 in raw beef and lamb products. *Epidemiology and Infection,* 2000 (April), 124 (2), 207-13.

26. Palumbo S.A., Call J.E., Schultz F.J., Williams A.C. Minimum and maximum temperatures for growth and verotoxin production by hemorrhagic strains of *Escherichia coli. Journal of Food Protection,* 1995, 58 (4), 352-6.

27. Meng J., Doyle M.P., Zhao T., Zhao S. Detection and control of *Escherichia coli* O157:H7 in foods. *Trends in Food Science and Technology,* 1994, 5 (6), 179-85.

28. Department of Health, MAFF. Safer cooked meat production guidelines. A 10-point plan. Heywood, Lancashire. BAPS Health Publications Unit, 1992.

29. Buchanan R.L., Bagi L.K. Expansion of response surface models for the growth of *Escherichia coli* O157:H7 to include sodium nitrite as a variable. *International Journal of Food Microbiology,* 1994, 23 (3+4), 317-32.

30. Glass K.A., Loeffelholz J.M., Ford J.P., Doyle M.P. Fate of *Escherichia coli* O157:H7 as affected by pH or sodium chloride and in fermented, dry sausage. *Applied and Environmental Microbiology,* 1992, 58 (9), 2513-6.

31. Park S., Worobo R.W., Durst R.A. *Escherichia coli* O157:H7 as an emerging foodborne pathogen: a literature review. *Critical Reviews in Food Science and Nutrition,* 1999, 39 (6), 481-502.

32. Abdul-Raouf U.M., Beuchat L.R., Ammar M.S. Survival and growth of *Escherichia coli* O157:H7 on salad vegetables. *Applied and Environmental Microbiology,* 1993, 59 (7), 1999-2006.

33. International Commission on Microbiological Specifications for Foods. Intestinally pathogenic *Escherichia coli*, in Microorganisms in foods, volume 5: microbiological specifications of food pathogens. International Commission on Microbiological Specifications for Foods. London. Blackie, 1996, 126-40.

Further reading

Institute of Food Science and Technology. Verocytotoxin-producing *E coli* Food Poisoning and its Prevention.. IFST, 2004.

Maunsell B., Bolton D.J. Guidelines for Food Safety Management on Farms. Teagasc. The National Food Centre, 2004.

Meng J., Doyle M.P., Zhao T., Zhao S. Enterohemorrhagic *Escherichia coli*, in Food Microbiology: Fundamentals and Frontiers. Eds Doyle M.P., Beuchat L.R., Montville T.J. 2nd edition. Washington DC. ASM Press, 2001, 193-213.

Duffy G., Garvey P., McDowell D.A. Trumbull. Verocytotoxigenic *E. coli*. Food and Nutrition Press, 2001.

International Life Sciences Institute. Approach to the control of entero-haemorrhagic *Escherichia coli* (EHEC). Brussels. ILSI Europe, 2001.

Stewart C.S., Flint H.J. *Escherichia coli* O157 in farm animals. Wallingford. CABI Publishing, 1999.

Food Safety Authority of Ireland. The prevention of *E. coli* O157:H7 infection: a shared responsibility. Dublin. FSAI, 1999.

Kaper, J.B., O'Brien, A.D. *Escherichia coli* O157:H7 and other Shiga toxin-producing *E. coli* strains. Washington DC. ASM Press, 1998.

Bell C., Kyriakides A. *E. coli*: a practical approach to the organism and its control in foods. London. Blackie, 1998.

Desmarchelier P.M., Grau F.H.. *Escherichia coli*, in Foodborne Microorganisms of Public Health Significance. 5th Edn. Ed. Australian Institute of Food Science and Technology. North Sydney. AIFST, 1997, 231-64.

Takeda Y. Enterohaemorrhagic *Escherichia coli*, in Food Safety and Foodborne Diseases. Ed. World Health Organization. Geneva. WHO, 1997, 74-80.

The Pennington Group, Pennington, T.H. Report on the circumstances leading to the 1996 outbreak of infection with *E. coli* O157 in Central Scotland, the implications for food safety and the lessons to be learned. Edinburgh. Stationery Office, 1997.

International Commission on Microbiological Specifications for Foods. Intestinally pathogenic *Escherichia coli*, in Microorganisms in Foods, Volume 5: Microbiological Specifications of Food Pathogens. Ed. International Commission on Microbiological Specifications for Foods. London. Blackie, 1996, 126-40.

Methods of detection

British Standards Institution. Microbiology of food and animal feeding stuffs. Horizontal method for the detection of *Escherichia coli* O157. BS EN ISO 16654:2001. (Incorporating Corrigendum No. 1.). BSI, 2001.

Wilson I.G. Detection of *Escherichia coli* O157:H7 by immunomagnetic separation and multiplex polymerase chain reaction. Food Microbiology Protocols. Eds Spencer J.F.T., de Spencer A.L.R. Totowa. Humana Press, 2001, 85-94.

Public Health Laboratory Service. Detection of *Escherichia coli* O157 by immunomagnetic bead separation. PHLS Standard Methods for Food Products. Public Health Laboratory Service. London. PHLS, 1999.

Woody J.-M., Stevenson J.A., Wilson R.A., Knabel S.J. Comparison of the Difco EZ Coli Rapid Detection System and Petrifilm Test Kit-HEC for detection of *Escherichia coli* O157:H7 in fresh and frozen ground beef. *Journal of Food Protection,* 1998 (January), 61 (1), 110-2.

Entis P. Direct 24-hour presumptive enumeration of *Escherichia coli* O157:H7 in foods using hydrophobic grid membrance filter followed by serological confirmation: collaborative study. *Journal of AOAC International,* 1998 (March-April), 81 (2), 403-18.

Stephan R. A review of isolation and detection procedures for *Escherichia coli* O157 and other verotoxin-producing *E. coli* (VTEC). Mitteilungen aus dem Gebiete der Lebensmitteluntersuchung und Hygiene, 1997, 88 (6), 681-92.

Heuvelink A.E., Zwartkruis-Nahuis J.T.M., de Boer E. Evaluation of media and test kits for the detection and isolation of *Escherichia coli* O157 from minced beef. *Journal of Food Protection,* 1997 (July), 60 (7), 817-24.

Vernozy-Rozand C. Detection of *Escherichia coli* O157:H7 and other verocytotoxin-producing *E. coli* (VTEC) in food. *Journal of Applied Microbiology,* 1997 (May), 82 (5), 537-51.

Hitchins A.D., Feng P., Watkins W.D., Rippey S.R., Chandler L.A. *Escherichia coli* and the coliform bacteria, in Bacteriological Analytical Manual. 8th Edn. Ed. Food and Drug Administration. Gaithersburg. AOAC International, 1995.

YERSINIA

Yersina enterocolitica was initially named *Bacterium enterocoliticum*, and the genus *Yersinia* was then proposed in 1944. However, the allocation of *Yersinia* to the family Enterobacteriaceae was only established in 1964. *Yersinia* includes three well-established pathogens – *Yersinia pestis*, *Yersinia pseudotuberculosis* and *Yersinia enterocolitica* – and several non-pathogens (1,2). *Yersinia enterocolitica* is the most common cause of illness; however, *Y. pseudotuberculosis* may have been the cause of some outbreaks in Europe and Japan (3,4). This section only considers *Y. enterocolitica*.

The Organism *Yersinia enterocolitica*

Yersinia are Gram-negative, oxidase negative, catalase positive non-sporing rods (occasionally coccoid) (2). Like other members of this family, *Yersinia* is a facultative anaerobe that ferments glucose. They are non-motile at 35-37 °C, but motile at 22-25 °C; however, some human pathogens strains of serovar O:3 are non-motile at both temperatures (1,2).

There are 11 currently recognised *Yersinia* species but, apart from *Y. pestis* (causative agent of bubonic and pneumonic plague), the *Yersinia* species of most significance to human health is *Y. enterocolitica*. Within this species, only certain strains that carry a virulence plasmid are generally considered to be pathogenic. If this plasmid is lost, this renders the strain non-virulent; virulence is also temperature-dependent (3,5,6). Elements encoded by the chromosome are also necessary for maximum virulence. Other pathogenic factors of *Yersinia* species include an enterotoxin and the capacity for cellular invasion (2,3,7).

There are many serotypes amongst the species *Y. enterocolitica*. Serotypes O:3 and O:9 and, to a lesser extent, O:5,27 are the predominant pathogenic serotypes in the UK and Europe. In the USA, serotype O:8 has been the most commonly reported cause of outbreaks of yersiniosis followed by O:5,27, but has now been overtaken by O:3, currently the most common serotype worldwide (2,8).

Yersinia Food Poisoning

Y. enterocolitica has only been recognised as a foodborne pathogen since about the mid-1970s. Yersiniosis is the gastrointestinal illness caused by the consumption of food containing viable pathogenic *Yersinia*.

Many strains of *Yersinia*, including serotypes O:3, O:8 and O:9, are capable of producing a heat-stable enterotoxin. However, this toxin may not play an important part in inducing yersiniosis.

Incubation time

The incubation period may range from 1-11 days (2,5). The illness is usually short-lived, but it can persist for 1-3 weeks (9), or even several months if there are complications.

Symptoms

Yersinia can cause a wide range of clinical symptoms, depending on the strain involved, the dose and the age and susceptibilty of the host. However, acute gastroenteritis - diarrhoea and abdominal pain - is the most common manifestation of infection.

Babies and children under five are most susceptible to yersiniosis. In babies, children and adolescents, the main symptoms are gastroenteritis and inflammation of the lymph glands; in adults, the predominant symptoms are abdominal disorders and diarrhoea. Complications include arthritis and skin disorders. Pharyngitis is also common in all age groups (2,6,7).

Intestinal pain from yersiniosis, especially amongst adolescents and adults, can be so severe as to be confused with appendicitis.

Fever - and less commonly, vomiting - can also occur.

Mortality

Death is rare, but can be associated with septic infections amongst persons with underlying disease.

Infective dose

The severity of symptoms is thought to be dose-related; however, the actual infective dose is not known. Certain reports suggest that it might be very high in adults, probably $>10^4$ cells (7), but lower in infants and the immunocompromised.

Foods involved

The most frequently reported mode of transmission of yersiniosis has been foods and water. However, the exact role of food in yersiniosis remains unclear.

In most countries milk appears to have acted as a vehicle of yersiniosis more often than any other food. Two outbreaks in Canada in the mid-1970s were associated with the consumption of raw milk. In the US, an outbreak during 1976, which affected 220 grade school children, was linked to the consumption of chocolate milk (3). Reconstituted milk and pasteurised milk have also been identified as the cause of other outbreaks in USA.

The main exception to milk as the vehicle of infection is in countries where pork is commonly eaten raw, such as in Belgium, where yersiniosis is relatively common (10,11).

Other foods implicated as sources in outbreaks include soyabean curd (tofu) packed in untreated spring water, and bean sprouts (2).

During the period 1986-88, there were a total of 10 laboratory-reported suspected incidents of foodborne yersiniosis in England and Wales. One – a family outbreak – was associated with the consumption of pork; the other sporadic cases were linked with milk (12).

Incidence of *Yersinia* Food Poisoning

The incidence of yersiniosis is not known, but it is often reported in the cooler regions of Europe and North America (1,2,13). It is of high prevalence in Belgium and the third most common bacterial infection in Sweden (6). In certain countries, such as Canada, Holland and Australia, *Y. enterocolitica* has surpassed *Shigella* as a cause of foodborne disease (8,13,14).

Laboratory reports of *Y. enterocolitica* infection increased in England and Wales from 45 in 1980 to about 570 in 1989. The reported incidence also increased in Scotland. Since 1989, numbers have decreased in England and

Wales from under 700 in 1991 to 12 (provisional data) in 2006. During the period 1986-88, there were only 10 incidents that were suspected as being foodborne. As with other forms of foodborne illness, the true incidence of infection is likely to be considerably higher than the reported incidence suggests (10).

Sources

Humans

Strains of *Y. enterocolitica* (mostly non-pathogenic) may be carried in low numbers as part of the transient intestinal flora of apparently healthy humans. A food handler was implicated in one outbreak of yersiniosis in the USA. Pathogenic strains have also been isolated from apparently healthy food handlers in Japan (15).

Person-to-person transmission, by the faecal-oral route, of serotypes O:3 and O:9 of *Y. enterocolitica* has been reported .

Animals and environment

The organism *Y. enterocolitica* is ubiquitous; it is present in a wide range of animals, especially pigs. The organism is present in the oral cavity, particularly the tongue and tonsils, lymph nodes, in the intestine and faeces (2). Environmental sources, including water, may act as a direct or indirect vehicle for *Yersinia* spp. However, it is important to note that most of these 'environmental' strains are not pathogenic.

It has been concluded that no food-producing animal other than pigs appears to be a common reservoir of pathogenic *Y. enterocolitica* (the organism being harboured mainly in the throat/tonsil area) (4,7,16).

Spread of yersiniosis by handling infected pigs, dogs, cats and rodents has been reported. Fleas may also be implicated (6).

Foods

The role of foods in the transmission of yersiniosis is not clearly understood, because most isolations from foods are through surveys, and not in response to an outbreak, so the isolates do not correspond with those from human illness. Foods found to be contaminated with *Y. enterocolitica* include milk and non-ripened/non-fermented dairy products, various meats and poultry, seafoods, vegetables and miscellaneous prepared foods, including salads. Milk and pork have received most attention in surveys because of their known association with outbreaks (2,4,17,18,19,20).

Growth/Survival Characteristics of the Organism in Foods

Temperature

Yersinias are psychrotrophic organisms, being capable of growth at refrigeration temperatures; doubling times at 10 °C are reported as 5 h. and, at 1 °C, approximately 40 h. Extremely slow growth has been recorded at temperatures as low as 0 °C to -1.3 °C. However, the optimum temperature for growth of *Y. enterocolitica* is 28-29 °C with the reported growth range of -2-42 °C (1,2,5). The maximum temperature where growth has been recorded is 44 °C (3,5).

The organism is quite resistant to freezing and has been reported to survive in frozen foods for extended periods (1,2).

Heat resistance

The organism is sensitive to heat, being easily killed at temperatures above about 60 °C. D-values determined in scaling water were 96, 27, and 11 seconds at 58 °C, 60 °C and 62 °C, respectively (2). Internal temperatures of 60 °C in beef roasts inactivate up to a million cells of *Yersinia* per gram, whereas 51 °C leaves some survivors (21). Therefore, it should be destroyed by milk pasteurisation and other commonly applied heat processes, such as cooking, boiling, baking and frying temperature (1).

pH

Yersinia enterocolitica is sensitive to pH values of less than 4.6 (more typically 5.0) in the presence of organic acids, e.g. acetic acid. *Y. enterocolitica* are not able to grow at pH <4.2 or >9.0. (A lower pH minimum for growth (pH 4.1-4.4) has been observed with inorganic acids, under otherwise optimal conditions.) Its optimum is pH 7.0-8.0.

Water activity/salt

Yersinia may grow at salt concentrations up to about 5% (*ca.* a_w 0.96), but no growth occurs at 7% (a_w <0.945). Growth is slowed in foods containing 5% salt (5,7).

Atmosphere

Carbon dioxide exhibits some inhibitory effect on the growth of *Y. enterocolitica*. Vacuum packaging can slow its growth to a lesser extent. Growth has been demonstrated in vacuum-packed and MAP meats (2,22).

Irradiation

Y. enterocolitica is among the most sensitive bacteria, needing the lowest radiation dose for elimination (D_{10} ~0.20kGy) (2).

Summary of Control of *Yersinia enterocolitica* in Foods

Because refrigeration is relatively ineffective in preventing the growth of *Y. enterocolitica* in foods, measures should be taken to minimise contamination with this organism in foods, especially those that are chilled. Appropriate heat treatments, where possible, should be carried out to ensure the destruction of the organism. Raw pork represents a particular hazard and should be treated as such. Special care should be taken when handling the carcass to prevent cross-contamination.

Additional note:

The mere presence of *Y. enterocolitica* in foods cannot give any real indication of its ability to cause illness; the organism would need to be tested further before there could be any certainty of its public health significance. Nevertheless, its presence in cooked foods indicates inadequate heat processing or post-process contamination and cannot be justified.

Bibliography

References

1 Nesbakeen T. *Yersinia enterocolitica*, in *Foodborne Infections and Intoxications*. Eds Reimann H.P., Cliver D.O. Oxford. Elsevier, 2006, 289-312.

2. Nesbakeen T. *Yersinia enterocolitica*, in *Emerging Foodborne Pathogens*. Eds Motarjemi Y., AdamsM. Cambridge. Woodhead Publishing Ltd., 2006, 373-405.

3. Feng P., Weagant S.D. *Yersinia*, in *Foodborne Disease Handbook, Vol. 1. Diseases Caused by Bacteria*. Eds Hui Y.H., Gorham J.R., Murrell K.D., Cliver D.O. New York. Marcel Dekker, 1994, 427-60.

4. Schiemann D.A. *Yersinia enterocolitica* and *Yersinia pseudotuberculosis*, in *Foodborne Bacterial Pathogens*. Ed. Doyle M.P. New York. Marcel Dekker, 1989, 601-72.

5. International Commission on Microbiological Specifications for Foods. *Yersinia enterocolitica*, in Microorganisms in foods, Vol. 5. Microbiological specifications of food pathogens. International Commission on Microbiological Specifications for Foods. London. Blackie, 1996, 458-78.

6. Sutherland J., Varnam A. Enterotoxin-producing *Staphylococcus*, *Shigella*, *Yersinia*, *Vibrio*, *Aeromonas* and *Pleisomonas*, in *Foodborne pathogens – Hazards, risk analysis and control.* Blackburn C. de W., McClure P.J. Cambridge. Woodhead Publications Limited., 2002, 390-5.

7. Robins-Browne R.M. *Yersinia enterocolitica*, in *Food Microbiology: Fundamentals and Frontiers*. Eds Doyle M.P., Beuchat L.R., Montville T.J. Washington DC. ASM Press, 1997, 192-215.

8. Barton M.D., Kolega V., Fenwick S.G. *Yersinia enterocolitica*. Foodborne Microorganisms of Public Health Significance. Ed. Australian Institute of Food Science and Technology. 5th edition. North Sydney: AIFST, 1997, 493-519.

9. Management of Outbreaks of Foodborne Illness. *Yersinia enterocolitica*. Ed. Department of Health. London. HMSO. 1994, 97-8.

10. Carniel E., Mollaret H.H. Yersiniosis (Review). *Comp. Immun. Microbiol. Infect. Dis*. 1990, 13 (2), 51-8.

11. Tauxe R.V., Vandepitte J., Wauters G., Martin S.M., Goossens U., de Mol P., van Noyen R., Thiers G. *Yersinia enterocolitica* infections and pork: the missing link. *Lancet,* 1987, 1 (8542), 1129-32.

12. Halligan A.C. The emerging pathogens - *Yersinia*, *Aeromonas* and verotoxigenic *E. coli* (VTEC) - a literature survey. *Leatherhead Food Research Association,* 1990, 73.

13. Doyle M.P. Pathogenic *Escherichia coli*, *Yersinia enterocolitica* and *Vibrio parahaemolyticus*. *Lancet,* 1990, 336 (8723), 1111-5.

14. Cover T.L., Aber R.C. *Yersinia enterocolitica*. *New England Journal of Medicine,* 1989, 321, 16-24.

15. Morse D.L., Shayegani M., Gallo R.J. Epidemiological investigation of a *Yersinia* camp outbreak linked to a food handler. *American Journal of Public Health,* 1984, 74 (6), 589-92.

16. Andersen J.K., Sorensen R., Glensbjerg M. Aspects of the epidemiology of *Yersinia enterocolitica*: a review. *International Journal of Food Microbiology*, 1991, 13 (3), 231-8.

17. Robins-Browne R.M., Hartland E.L. *Yersinia* Species, in *International Handbook of Foodborne Pathogens.* Eds Miliotis M.D., Bier J.W. New York. Marcel Dekker Inc., 2003, 323-56.

18. Schiemann D.A. *Yersinia enterocolitica* in milk and dairy products. *Journal of Dairy Science,* 1987, 70 (2), 383-91.

19. Greenwood M.H., Hooper W.L. *Yersinia* spp. in foods and related environments. *Food Microbiology,* 1985, 2 (4), 263-9.

20. Kapperud G. *Yersinia enterocolitica* in food hygiene. *International Journal of Food Microbiology,* 1991, 12 (1), 53-65.

21. Hanna M.O., Stewart J.C., Carpenter Z.L., Vanderzant C. Effect of heating, freezing and pH on *Yersinia enterocolitica*-like organisms from meat. *Journal of Food Protection,* 1997, 40 (10), 689-92.

22. Doherty A., Sheridan J.J., Allen P., McDowell D.A., Blair I.S., Harrington D. Growth of *Yersinia enterocolitica* O:3 on modified atmosphere packaged lamb. *Food Microbiology,* 1995, 12 (3), 251-7.

Further reading

Robins-Browne R.M. *Yersinia enterocolitica*, in *Food Microbiology: Fundamentals and Frontiers.* Eds Doyle M.P., Beuchat L.R., Montville T.J. 2nd edition. Washington DC. ASM Press, 2001, 215-45.

Minnich S.A., Smith M.J., Weagant S.D., Feng P. *Yersinia*, in *Foodborne Disease Handbook, Vol. 1.: Bacterial Pathogens.* Eds Hui Y.H., Pierson M.D., Gorham J.R. 2nd edition. New York. Marcel Dekker, 2000, 471-514.

Bodnaruk P.W., Draughon F.A. Effect of packaging atmosphere and pH on the virulence and growth of *Yersinia enterocolitica* on pork stored at 4 °C. *Food Microbiology,* 1998 (April), 15(2), 129-36.

Various authors. *Yersinia enterocolitica.* Food Associated Pathogens: Proceedings of a Symposium. Uppsala, May 1996. Ed. International Union of Food Science and Technology. Uppsala. SLU, 1996, 198-206.

Catteau M. The genus *Yersinia*, in *Microbiological Control for Foods and Agricultural Products.* Eds Bourgeois C.M., Leveau J.Y. Cambridge. VCH Publishers, 1995, 335-45.

Toora S., Budu-Amoaka E., Ablett R.F., Smith J. Effect of high-temperature short-time pasteurization, freezing and thawing and constant freezing, on the survival of *Yersinia enterocolitica* in milk. *Journal of Food Protection,* 1992, 55 (10), 803-5

Brackett R.E. Effects of various acids on growth and survival of *Yersinia enterocolitica. Journal of Food Protection*, 1987, 50(7)589-601+607.

Methods of detection

Lambertz S.T., Granath K., Fredriksson-Ahomaa M., Johansson K.-E., Danielsson-Tham M.-L. Evaluation of a combined culture and PCR method (NMKL-163A) for detection of presumptive pathogenic *Yersinia enterocolitica* in pork products. *Journal of Food Protection*, 2007, (February), 70 (2), 335-40

Cocolin L, Comi G. Use of a culture-independent molecular method to study the ecology of *Yersinia* spp in food. *International Journal of Food Microbiology*, 2006, (November 15), 105 (1), 71-82

Lambertz S.T., Danielsson-Tham M.-L. Identification and characterization of pathogenic *Yersinia enterocolitica* isolates by PCR and pulsed-field gel electrophoresis. *Applied and Environmental Microbiology*, 2005, (July), 71 (7), 3674-81.

Weagant S.D., Feng P. *Yersinia*, in *Compendium of methods for the microbiological examination of foods.* American Public Health Association, Downes F.P., Ito K. 4th edition. Washington DC. APHA, 2001, 421-8.

Thisted Lambertz S., Lindqvist R., Ballagi-Pordany A., Danielsson-Tham M.L. A combined culture and PCR method for detection of pathogenic *Yersinia enterocolitica* in food. *International Journal of Food Microbiology,* 2000, (June 10), 57 (1-2), 63-73.

Lambertz S.T., Ballagi-Pordany A., Nilsson A., Norberg P., Danielsson-Tham M.-L. A comparison between a PCR method and a conventional culture method for detecting pathogenic *Yersinia enterocolitica* in food. *Journal of Applied Bacteriology,* 1996, 81 (3), 303-8.

Weagant S.D., Feng P., Stanfield J.T. *Yersinia enterocolitica* and *Yersinia pseudotuberculosis*, in *Bacteriological Analytical Manual.* Ed. Food and Drug Administration. 8th edition. Gaithersburg. AOAC International, 1995, 13.

de Boer E. Isolation of *Yersinia enterocolitica* from foods. Culture Media for Food Microbiology. Eds Corry J.E.L., Curtis G.D.W., Baird R.M. Amsterdam. Elsevier, 1995, 219-28.

Curtis L.M., Blackburn C. de W. Evaluation of four cultural methods for the detection of *Yersinia enterocolitica* in foods. *Journal of Applied Bacteriology,* 1995, 79 (1), supplement, 16.

Landgraf M., Iaria S.T., Falcao D.P. An improved enrichment procedure for the isolation of *Yersinia enterocolitica* and related species from milk. *Journal of Food Protection,* 1993, 56 (5), 447-50.

OTHER BACTERIA THAT MAY BE FOODBORNE AND HAVE HEALTH IMPLICATIONS

SHIGELLA

Shigellas are Gram-negative, non-motile, oxidase-negative, catalase-positive rods (although with some exceptions), belonging to the family Enterobacteriaceae, and are closely related to *E. coli*. There are a number of species involved in the disease shigellosis, but *S. sonnei* accounts for the greatest proportion of cases in developed countries. *Shigella flexneri*, *S. boydii* and *S. dysenteriae* are also important species (1,2).

Foodborne shigellosis is rare in the UK. Shigellosis is a human disease; it is host-adapted to humans, and its transmission is mainly via person-to-person contact by the faecal-oral route. It is not indigenous in foods. However, water, milk and occasionally food can be a vehicle for the organism where hygiene is inadequate, usually in developing countries, and especially amongst children, particularly under six years of age. In developed countries, transmission can occur through infected food handlers, and it is a major cause of illness in the USA. Most foodborne incidents of shigellosis involve poor personal hygiene amongst handlers, especially the failure to observe hand-washing requirements after using the toilet before handling food (1,3,4).

Most foods involved in foodborne shigellosis (in the USA) are salads; however, a number of other food types have been implicated, most involving infected food handlers (1,3,4).

In 1994, there was an outbreak of *Shigella sonnei* infection traced to iceberg lettuce imported from Spain, and cases of the illness were detected in several European countries, including the UK (5,6). The lettuces had been contaminated with faecal material, but the cause of the contamination is unknown (6). There has also been a recent *Shigella flexneri* outbreak in the UK associated with fruit salad obtained from a supermarket salad bar (7). Another outbreak occurred with fresh parsley in the US and Canada, traced back to a farm in Mexico (8).

Shigellosis can range from an asymptomatic infection (asymptomatic carriers can be a source of the organism) to mild diarrhoea, to dysentery. Blood and/or mucus and pus may be apparent in the stools in severe cases, and other symptoms can include dehydration, vomiting and fever. Shigellosis is highly infectious, involving only a very small infective dose (possibly as few as about 10 cells) (1). The incubation period can range from 12 h to 7 days, usually 1-3 days. The illness persists for up to 2 weeks.

Shigella is a relatively fragile organism; it is easily destroyed by heat (e.g. 63 °C for 5 min) (1); some strains can grow at temperatures between 7 °C and 46 °C (the optimum is 37 °C); its reported pH range for growth is about pH 5-8 (depending on acid type, etc.), but *Shigella* will not survive for long in acid foods, although there is some evidence that it can survive 13-92 days in mayonnaise-containing salads, and some cheese products (9). The organism dies only slowly at reduced a_w and survives in both frozen and chilled foods (10).

Proper attention to personal hygiene, as well as the control of temperature of foods, will prevent the contamination and growth of *Shigella* in foods.

AEROMONAS

Aeromonas species are Gram-negative, oxidase and catalase positive, facultatively anaerobic rods, of variable motility, belonging to the family Aeromonadaceae, although formally within the Vibrionaceae, like *Plesiomonas*. There is still some dispute over the placing of this genus. While it is well known that certain aeromonads can cause infections in both fish and man, the role of *Aeromonas* spp in foodborne illness remains unclear. However, since the 1970s there has been increasing evidence implicating *Aeromonas* spp in incidents of human gastroenteritis, and a few incidents indicating a food vehicle (11,12).

Its purported involvement as a foodborne pathogen is based on several factors: its isolation in faeces from patients suffering food poisoning in the absence of any other pathogen, an immunological response after infection, work on pathogenicity in laboratory animals, and its presence in the types of foods commonly associated with food poisoning outbreaks.

The three *Aeromonas* species of potential significance in relation to food safety are the motile species of *Aeromonas*: *A. hydrophila*, *A. caviae* and *A. sobria* (11,12).

The most common type of gastroenteritis involving *Aeromonas* is a cholera-like illness (watery stools and mild fever); less common is a dysentery-like illness (blood and mucus in the stools). The illness is usually mild and self-limiting, but can be severe, especially in children, those over 50 years of age, and the immunocompromised.

The ability of a given strain of *Aeromonas* to cause illness appears to be linked with its ability to produce a number of 'virulence factors'. These include relatively heat-sensitive enterotoxins, which may or may not need to be pre-formed in food to cause illness (not yet clearly established), β-haemolysin, haemagglutinins and other factors. The ability of aeromonads to produce these virulence factors also appears to be linked with temperature; isolates from chilled foods or water may not be capable of growth or the production of virulence factors at body temperature. Thus foodborne isolates may or may not be capable of causing infection (11,12,13).

The main source of *Aeromonas* species is generally accepted to be water (hence the name, *A. hydrophila* - water-loving), including chlorinated and potable water. There have been a number of well-documented waterborne outbreaks, and water may contribute to the contamination of foods with *Aeromonas*. The organism is also widely distributed in nature, in a wide range of animals, including cows, and is occasionally carried by humans (11,12,13).

Aeromonas is a very common contaminant in foods, including seafoods, meat and poultry, milk, vegetables and salads. Numbers can occasionally exceed 10^5/g. It would seem, therefore, that *Aeromonas* is commonly consumed with food, with no apparent detrimental effect on health. There are only a few reports of 'food poisoning' from *Aeromonas* in foods, and these have involved mainly prawns or oysters (3,12).

The infective dose for *Aeromonas* has yet to be established, although it is likely to be relatively high, given the information above.

Aeromonads vary in their ability to grow at different temperatures. Potentially, the organism has a wide temperature range for growth and many strains are psychrotrophic, being able to grow at 5 °C and some as low as 2 °C (11,12,13), with an optimum at about 28 °C (14). *Aeromonas* is very heat-sensitive, being readily destroyed by pasteurisation or equivalent heat treatments. At 48 °C, D-values for different isolates heated in raw milk ranged from 3.20-6.23 min (11). Under otherwise optimal conditions, it has potential to grow over a pH range of *ca* 4-10 (12), and most strains are sensitive to >4.5% NaCl (11).

PLESIOMONAS SHIGELLOIDES

Plesiomonas is a facultatively anaerobic Gram-negative, catalase- and oxidase-positive, mostly motile rod belonging to the family Vibrionaceae, although very closely related to *Aeromonas*. A small number of infections among humans have been associated with this organism, mainly involving raw oysters. It is also naturally present in fresh and marine waters and as part of the natural flora of finfish and shellfish, especially in warmer countries. It has been isolated at low rates (<1%) from healthy humans in Japan, and from various mammals, including cattle and pigs, as well as from birds (3,15,16).

Symptoms associated with *Plesiomonas shigelloides* gastroenteritis include diarrhoea, abdominal pain and nausea, with fever, headache and vomiting being less common. The symptoms usually appear within 48 h, and the duration is usually 1-9 days (15,17).

The major distinction, in practical terms, between *Aeromonas* and *Plesiomonas* is growth temperature; *Plesiomonas* is not a psychrotroph - some strains can grow at 8 °C, but not at 5 °C. (The upper temperature limit for growth appears to be about 45 °C.) Adequate refrigeration, therefore, is effective in limiting the growth of *Plesiomonas* in foods. The organism is also heat-sensitive, being unable to survive heating at 60 °C for 30 min (15).

The infectious dose is unknown, but presumed to be $>10^6$ cells (18).

Other characteristics of interest include the ability of both clinical and environmental isolates of *Plesiomonas* to grow in 4% NaCl and at pH 4.5-8.5 (15,16).

PSEUDOMONAS AERUGINOSA

Pseudomonads are Gram-negative, motile, catalase-positive, oxidase-positive or negative, strictly aerobic rods, and members of the family Enterobacteriaceae. *Pseudomonas aeruginosa* has been shown to be capable of producing enterotoxins and has, infrequently, been linked with food poisoning. *Ps. cocovenenans* (recently reclassified as *Burkholdaria cocovenenans*) can also produce toxic substances, and has been associated with hypoglycaemia and death from the Indonesian food tempeh bongkrek (19,20,21,22).

Ps. aeruginosa is a classic example of an opportunistic pathogen. It rarely causes problems in healthy individuals; however, in infants, hospital patients and immunocompromised persons, the organism can cause a range of serious illnesses. Thus, infection from *Ps. aeruginosa* can be of particular concern in hospitals and, in infants, profuse diarrhoea can lead to death. Illness from *Ps. aeruginosa* in otherwise healthy adults takes the form of a mild enteritis. Intestinal illness from *Ps. aeruginosa* appears to involve a large infective dose ($>10^6$) (19).

The main route of transmission is person-to-person, but *Ps. aeruginosa* can occasionally be transmitted through water or food.

Pseudomonads are ubiquitous in soil, animals, plants and water - their natural habitat is soil, water and sewage - and several surveys carried out in hospitals have indicated contamination of vegetables, meats and frozen foods with *Ps. aeruginosa* (19).

Pseudomonads are sensitive to heat and drying but are resistant to disinfectants (19).

It has been concluded that the presence of *Ps. aeruginosa* in foods does not justify specific attention for healthy people.

OTHER MEMBERS OF THE ENTEROBACTERIACEAE

There are a number of genera besides those better-known members of the Enterobacteriaceae such as *Salmonella* and *E. coli* that may cause 'opportunistic' gastroenteritis when ingested in large numbers. These include *Edwardsiella tarda*, *Klebsiella* spp and *Proteus*. *Providencia* and *Enterobacter* have also been inconclusively suggested as being involved in foodborne enteritis (19,20), though there is increasing evidence that they may occasionally cause food poisoning (18).

A severe outbreak of gastroenteritis, followed by haemolytic uraemic syndrome, was associated with the consumption of green butter (i.e. butter containing parsley) sandwiches. Isolates of identical verotoxigenic *Citrobacter freundii* were recovered from the parsley and patients' faecal samples (23). As *C. freundii* is widely distributed in the environment, this incident indicates the potential importance of this organism in causing foodborne illness.

STREPTOCOCCI/ENTEROCOCCI

The streptococci are Gram-positive, catalase negative, non-motile facultatively anaerobic coccoid-shaped bacteria. They are mainly harmless, but a few are pathogenic to humans. These pathogenic species are haemolytic and some produce toxins. Some of the pathogenic types may be foodborne and have been particularly associated with milk and dairy products. However, since the broad application of milk pasteurisation, hazards are largely restricted to places where milk is still consumed raw or to where there has been a failure in pasteurisation. Brief mention will be made of three types of streptococci that have raised or are currently raising questions in relation to food safety (19).

Streptococcus pyogenes

Str. pyogenes is the characteristic species of the so-called Lancefield Group A streptococci (GAS). Its natural reservoir is humans (although it has also been isolated from cases of mastitis in cows), and it is one of the most common bacterial agents associated with upper respiratory tract and skin infections.

Str. pyogenes is a common cause of pharyngitis, and can also cause septic sore throat, tonsilitis, and scarlet fever. Most illness is transmitted via person-to-person contact or through airborne spread. However, these non-gastrointestinal diseases can be carried to other individuals via food. In addition, some cases of gastroenteritis have been attributed to *Str. pyogenes*; symptoms may be dose-dependent (3,19,20).

An extremely large outbreak (involving over 10,000 cases and 200 deaths) occurred in the USA in 1912, when milk handlers suffering pharyngitis transmitted the organism to milk (24), which was not subsequently pasteurised because of equipment failure. Other foods involved in the transmission of scarlet fever or septic sore throat caused by *Str. pyogenes* or similar GAS include seafood and other salads, ice cream, custards, macaroni cheese and sandwiches. These infections commonly involved infected food handlers, as well as inadequate temperature control. There have also been associations made between skin infections, meat or poultry handling and *Str. pyogenes*.

In most situations, control against infection with *Str. pyogenes* can be assured by food handler hygiene, coupled with thorough cooking/heat treatments where appropriate and by rapid chilling of foods to below 7 °C.

Other streptococci

The Lancefield Group C organism *Str. zooepidemicus* - a cause of mastitis in cows - was reported to cause a milkborne outbreak in humans, involving seven deaths, in the UK in 1984. This occurred after consumption of contaminated unpasteurised milk (20).

Other types of streptococci have also been implicated in incidents of skin infections amongst meat handlers, as well as with various other extra-intestinal infections, which are infrequently linked with foods. A discussion of these bacteria and the involvement of foods in their transmission can be found in Stiles (19), and Hajmeer and Fung (18).

Enterococci

Probably more familiar to food microbiologists are the enterococci, otherwise known as Group D streptococci or faecal streptococci. These bacteria encompass a small number of species, including the species formerly known as *Str. faecalis* and *Str. faecium* (now known as *Enterococcus faecalis* or *E. faecium*) (3,19,20).

The enterococci are normally present in the faeces of mammals, including man, but may also be found in a wide range of environments, such that their presence in foods may not be a reliable indicator of direct faecal contamination. They are relatively heat-resistant, so they can be the only surviving bacteria (apart from spore-formers) in pasteurised foods. They are also relatively resistant to drying (19).

The role of enterococci in foodborne illness remains uncertain. It is recognised that enterococci can cause a variety of extra-intestinal infections, and they have been implicated in infant diarrhoea in developing countries. They have also been implicated in a small number of apparent foodborne outbreaks, where large numbers of enterococci have been reported in the absence of other better-known foodborne pathogens. In such cases, mild 'food-poisoning symptoms' (including nausea, pains and diarrhoea) have been reported. Results from human volunteer studies where large numbers of enterococci have been ingested are somewhat conflicting. It is possible that certain strains are pathogenic under certain conditions, and more studies are needed to improve our understanding of the pathogenicity of foodborne isolates (3,18,19,20).

Enterococci can be found in a wide range of foods, but, as indicated, their significance in foods is not clear.

MYCOBACTERIUM AVIUM SUBSP. *PARATUBERCULOSIS*

It has recently been suggested that *Mycobacterium avium* subsp. *paratuberculosis* (MAP) may be a potential foodborne pathogen. Although it is normally associated with disease in cattle, there is a growing body of evidence that it may have a role in the development of a chronic inflammatory bowel condition called Crohn's Disease. In part, this is because Crohn's disease in humans has a very similar aetiology to Johne's disease in cattle, which is caused by MAP. The link is not currently proven, but concerns that MAP may survive milk pasteurisation because of the relatively high s of the organism have led to recommendations for extended high temperature/short time (HTST) milk pasteurisation treatments of 72 °C for 25 s instead of the standard 15 s (25,26,27), although this enhanced treatment may not completely destroy all strains of the organism.

Few data are available concerning the prevalence of MAP from environmental sources, although it has been isolated occasionally from water; milk still appears to be the main source of the organism. Very little work has been carried out on the survival of MAP from food. Ultimate control is by ensuring that the herd is free of the disease.

Bibliography

References

1. Doyle M.P. *Shigella*, in *Foodborne Diseases*. Ed. Cliver D.O. London. Academic Press, 1990, 205-8.

2. Roberts C., Cartwright R.Y. *Shigella sonnei* infection and its control. *PHLS Microbiology Digest,* 1993, 10 (3), 44-50.

3. Eley A.R. Other bacterial pathogens, in *Microbial Food Poisoning*. Ed. Eley A.R. 2nd edition. London. Chapman & Hall, 1996, 57-73.

4. Reed G.H. Foodborne illness (Part 12). Shigellosis. *Dairy, Food and Environmental Sanitation,* 1994, 14 (10), 591.

5. Anon. A foodborne outbreak of *Shigella sonnei* infection in Europe.*CDR Weekly,* 1994, 4 (25), 115.

6. Kapperud G., Rorvik L.M., Hasseltvedt V., Hoiby E.A., Iversen B.G., Staveland K., Johnsen G., Leitao J., Herikstad H., Andersson Y., Langeland G., Gondrosen B., Lassen J. Outbreak of *Shigella sonnei* infection traced to imported iceberg lettuce. *Journal of Clinical Microbiology,* 1995, 33 (3), 609-14.

7. Anon. An outbreak of infection with *Shigella flexneri* 1b in South East England. *CDR,* 1998, 8 (34), 297 + 300.

8. Centers for Disease Control and Prevention. Incidence of foodborne illness: preliminary data for the Foodborne diseases Active Surveillance Network, (FoodNet). United States, 1998, *Morbidity and Mortality Weekly Report* 48, 1999, 189-94.

9. Lampel K.A. *Shigella* species, in *Foodborne Pathogens, Microbiology and Molecular Biology.* Eds Fratamico P.M., Bhunia A.K., Smith J.L. Wymondham. Caister Academic Press, 2005, 341-56.

10. International Commission on Microbiological Specifications for Foods. *Shigella*, in Microorganisms in Foods, Vol. 5. Microbiological Specifications of Food Pathogens. Ed. International Commission on Microbiological Specifications for Foods. London. Blackie, 1996, 280-98.

11. Abeyta C., Palumbo S.A., Stelma G.N. *Aeromonas hydrophila* group, in *Foodborne Disease Handbook, Vol. 1. Diseases Caused by Bacteria.* Eds Hui Y.H., Gorham J.R., Murrell K.D., Cliver D.O. New York. Marcel Dekker, 1994, 1-27.

12. Stelma G.N. *Aeromonas hydrophila*, in *Foodborne Bacterial Pathogens.* Ed. Doyle M.P. New York. Marcel Dekker, 1989, 1-19.

13. Kirov S.M. The public health significance of *Aeromonas* spp. in foods. *International Journal of Food Microbiology,* 1993, 20 (4), 179-98.

14. Kovacek K., Faris A. *Aeromonas* species, in *International Handbook of Foodborne Pathogens.* Eds Miliotis M.D., Bier J. W. New York. Marcel Dekker Inc., 357-67.

15. Koburger J.A. *Plesiomonas shigelloides*, in *Foodborne Bacterial Pathogens* Ed. Doyle M.P. New York. Marcel Dekker, 1989, 311-25.

16. Varnam A.H., Evans M.G. *Plesiomonas shigelloides*, in *Foodborne Pathogens: An Illustrated Text.* Eds Varnam A.H., Evans M.G. London. Wolfe Publishing Ltd., 1991, 201-8.

17. International Commission on Microbiological Specifications for Foods. *Plesiomonas*, in Microorganisms in Foods, Vol. 5. Microbiological Specifications of Food Pathogens. Ed. International Commission on Microbiological Specifications for Foods. London. Blackie, 1996, 208-13.

18 Hajmeer M.N., Fung D.Y.C. Infections with Other Bacteria, in *Foodborne Infections and Intoxications*, Riemann H.P., Oliver D.O. 3rd Edition. Academic Press, 2006, 365.

19. Stiles M.E. Less recognized or presumptive foodborne pathogenic bacteria, in *Foodborne Bacterial Pathogens.* Ed. Doyle M.P. New York. Marcel Dekker, 1989, 673-733.

20. Varnam A.H., Evans M.G. Other bacterial agents of foodborne disease, in *Foodborne Pathogens: An Illustrated Text.* Eds Varnam A.H., Evans M.G. London. Wolfe Publishing Ltd., 1991, 355-61.

21. Taylor S.L. Other microbial intoxications, in *Foodborne Diseases.* Ed. Cliver D.O. London. Academic Press, 1990, 159-70.

22. Cox J.M., Kartadarma E., Buckle K.A. *Burkholdaria cocovenenans*, in *Foodborne Microorganisms of Public Health Significance.* Ed. Australian Institute of Food Science and Technology. 5th edition. North Sydney. AIFST, 1997, 521-30.

23. Tschape H., Prager R., Streckel W., Fruth A., Tietze E., Bohme G. Verotoxinogenic *Citrobacter freundii* associated with severe gastroenteritis and cases of haemolytic uraemic syndrome in a nursery school: green butter as the infection source. *Epidemiology and Infection,* 1995, 114 (3), 441-50.

24. Johnson E.A. Infrequent microbial infections, in *Foodborne Diseases.* Ed. Cliver D.O. London. Academic Press, 1990, 259-73.

25. Hammer P., Knappstein K., Hahn G. Significance of *Mycobacterium paratuberculosis* in milk, in Bulletin of the International Dairy Federation, No. 330. Ed. International Dairy Federation. Brussels. IDF, 1998, 12-16.

26. Grant I.R., Ball H.J., Rowe M.T. Effect of high-temperature, short-time (HTST) pasteurisation on milk containing low numbers of *Mycobacterium paratuberculosis. Letters in Applied Microbiology,* 1998 (February), 26 (2), 166-70.

27. Sung N., Collins M.T. Thermal tolerance of *Mycobacterium paratuberculosis. Applied and Environmental Microbiology,* 1998 (March), 64 (3), 999-1005.

Further reading

Griffiths M. *Mycobacterium paratuberculosis*, in *Emerging Foodborne Pathogens*. Eds Motarjemi J., Adams M. Cambridge. Woodhead Publishing Ltd, 2006, 522-56.

Isonhood J.H., Drake M. *Aeromonas* species in foods. *Journal of Food Protection,* 2002 (March), 65 (3), 575-82.

Fernandez-Escartin E., Garcia S. Miscellaneous agents: *Brucella*, *Aeromonas*, *Plesiomonas*, and beta-hemolytic streptococci, in *Guide to Foodborne Pathogens.* Labbe R.G., Garcia S. New York. Wiley, 2001, 295-313.

Kirov S.M. *Aeromonas* and *Plesiomonas* species, in *Food Microbiology: Fundamentals and Frontiers*. Eds Doyle M.P., Beuchat L.R., Montville T.J. 2nd edition. Washington DC. ASM Press, 2001, 301-27.

Lampel K.A., Maurelli A.T. *Shigella* species, in *Food Microbiology: Fundamentals and Frontiers.* Eds Doyle M.P., Beuchat L.R., Montville T.J. 2nd edition. Washington DC. ASM Press, 2001, 247-61.

Lampel K.A., Madden J.M., Wachsmuth I.K. *Shigella* species, in *The Microbiological Safety and Quality of Food, Volume 2.* Eds Lund B.M., Baird-Parker T.C., Gould G.W. Gaithersburg. Aspen Publishers, 2000, 1300-16.

Palumbo S., Stelma G.N., Abeyta C. The *Aeromonas hydrophila* group, in *The Microbiological Safety and Quality of Food, Volume 2.* Eds Lund B.M., Baird-Parker T.C., Gould G.W. Gaithersburg. Aspen Publishers, 2000, 1011-28.

Stiles M.E. Less recognized and suspected foodborne bacterial pathogens, in *The Microbiological Safety and Quality of Food, Volume 2.* Eds Lund B.M., Baird-Parker T.C., Gould G.W. Gaithersburg. Aspen Publishers, 2000, 1394-1419.

Franz C.M.A.P., Holzapfel W.H., Stiles M.E. Enterococci at the crossroads of food safety? *International Journal of Food Microbiology,* 1999 (March 1), 47 (1-2), 1-24.

Holmes B., Aucken H.M. *Citrobacter*, *Enterobacter*, *Klebsiella*, *Serratia* and other members of the Enterobacteriaceae, in T*opley and Wilson's Microbiology and Microbial Infections, Volume 2: Systematic Bacteriology.* Eds Balows A., Duerden B.I. 9th edition. London. Arnold Publishers, 1998, 999-1033.

Kilian M. *Streptococcus* and *Lactobacillus*, in T*opley and Wilson's Microbiology and Microbial Infections, Volume 2: Systematic Bacteriology.* Eds Balows A., Duerden B.I. 9th edition. London. Arnold Publishers, 1998, 633-67.

Cox J.M., Kartadarma E., Buckle K.A. *Burkholdaria cocovenenans*, in Foodborne Microorganisms of Public Health Significance. Australian Institute of Food Science and Technology. 5th edition. North Sydney. AIFST, 1997, 521-30.

Nazarowec-White M., Farber J.M. *Enterobacter sakazakii*: a review. *International Journal of Food Microbiology,* 1997, 34 (2), 103-13.

International Commission on Microbiological Specifications for Foods. *Streptococcus*, in Microorganisms in Foods, Vol. 5. Microbiological Specifications of Food Pathogens. Ed. International Commission on Microbiological Specifications for Foods. London. Blackie, 1996, 334-46.

International Commission on Microbiological Specifications for Foods. *Shigella*, in Microorganisms in Foods, Vol. 5. Microbiological Specifications of Food Pathogens. Ed. International Commission on Microbiological Specifications for Foods. London. Blackie, 1996, 280-98.

International Commission on Microbiological Specifications for Foods. *Pseudomonas cocovenenans*, in Microorganisms in Foods, Vol. 5. Microbiological Specifications of Food Pathogens. Ed. International Commission on Microbiological Specifications for Foods. London. Blackie, 1996, 214-6.

Methods of detection

New FDA Microbiological Methods (proposed for inclusion in the BAM). Isolation and enumeration of *Enterobacter sakazakii* from dehydrated powdered infant formula. FDA Microbiological Methods. Food and Drug Administration, 2002. http://www.cfsan.fda.gov/~comm/mmesakaz.html

Hartman P.A., Deibel R.H., Sieverding L.M. Enterococci, in Compendium of Methods for the Microbiological Examination of Foods. American Public Health Association, Downes F.P., Ito K. 4th edition. Washington DC. APHA, 2001, 83-7.

Palumbo S., Abeyta C., Stelma G., Wesley I.W., Wei C.-I., Koberger J.A., Franklin S.K., Schroeder-Tucker L., Murano E.A. *Aeromonas*, *Arcobacter*, and *Plesiomonas*, in Compendium of Methods for the Microbiological Examination of Foods. American Public Health Association, Downes F.P., Ito K. 4th edition. Washington DC. APHA, 2001, 283-300.

Lampel K.A. *Shigella*, in Compendium of Methods for the Microbiological Examination of Foods. American Public Health Association, Downes F.P., Ito K. 4th edition. Washington DC. APHA, 2001, 381-5.

Grant I.R., Rowe M.T. Methods of detection and enumeration of viable *Mycobacterium paratuberculosis* from milk and milk products, in *Mycobacterium paratuberculosis*. International Dairy Federation. Brussels. IDF, 2000, 41-52, No. 363.

Andrews W.H., June G.A., Sherrod P. *Shigella*, in Bacteriological Analytical Manual. Ed. Food and Drug Administration. 8th edition. Gaithersburg: AOAC International, 1995, 6.

Jeppesen C. Media for *Aeromonas* spp., *Plesiomonas shigelloides* and *Pseudomonas* spp. from food and environment, in *Culture Media for Food Microbiology.* Eds Corry J.E.L., Curtis G.D.W., Baird R.M. Amsterdam. Elsevier, 1995, 111-27.

Reuter G. Culture media for enterococci and group D-streptococci, in *Culture Media for Food Microbiology*. Eds Corry J.E.L., Curtis G.D.W., Baird R.M. Amsterdam. Elsevier, 1995, 51-61.

Collins C.H., Lyne P.M., Grange J.M. *Proteus*, *Providencia* and *Morganella*, in *Collins and Lyne's Microbiological Methods*. Eds Collins C.H., Lyne P.M., Grange J.M. 7th edition. Oxford. Butterworth-Heinemann Ltd, 1995, 326-7.

Collins C.H., Lyne P.M., Grange J.M. *Escherichia*, *Citrobacter*, *Klebsiella* and *Enterobacter*, in *Collins and Lyne's Microbiological Methods*. Eds Collins C.H., Lyne P.M., Grange J.M. 7th edition. Oxford. Butterworth-Heinemann Ltd, 1995, 305-11.

Anon. MAFF validated methods for the analysis of foodstuffs. Method for the detection of *Pseudomonas aeruginosa* in natural mineral waters by liquid enrichment. *Journal of the Association of Public Analysts,* 1993, 29 (4), 279-84.

Food and Agriculture Organization, Andrews, W. *Shigella*, in Manual of Food Quality Control, Vol. 4. Microbiological Analysis. Eds Food and Agriculture Organization, Andrews W. Rome. FAO, 1992, 49-56.

Grover S., Batish V.K., Srinivasan R.A. Production and properties of crude enterotoxin of *Pseudomonas aeruginosa. International Journal of Food Microbiology,* 1990, 10 (3/4), 201-8.

FOODBORNE VIRUSES AND PROTOZOA

VIRUSES

Viruses are an important cause of gastroenteritis and, although most illness is caused by person-to-person spread, food plays a part in the transmission of viral gastroenteritis. Many of the incidents of food poisoning that are of 'unknown aetiology' (where no food-poisoning bacteria are detected) are likely to be caused by viruses (1). However, the actual role of viruses in the incidence of 'food poisoning' is little understood, mainly because it is virtually impossible to detect foodborne viruses with current techniques.

Viral gastroenteritis was first recognised in a school outbreak in Norwalk, Ohio, in 1969. The virus was detected in stool samples and became known as the 'Norwalk agent'. Since this time, the role of viruses in foodborne illness has become increasingly recognised, although most cases probably go unreported and the incidence of foodborne transmission of illness remains poorly understood (2).

The first confirmed outbreaks of foodborne viral gastroenteritis in the UK occurred in the south of England in 1976 and 1977 from contaminated cockles. This was followed by an outbreak of hepatitis A in 1978 in the Midlands and north of England attributed to mussels. In all incidents, the shellfish had been harvested from sewage-polluted waters and were inadequately cooked. Also, viral gastroenteritis arising from the consumption of oysters has been a major problem (2).

The viruses that are known to be important to food safety (i.e. they are known to be foodborne) fall mainly into two distinct groups: 1) the group that causes gastroenteritis – now correctly termed noroviruses (previously referred to as Small Round Structured Viruses (SRSVs), the Norwalk group, or Norwalk-like viruses (NLV)); and 2) hepatitis A virus (HAV) - the cause of 'infectious hepatitis' or viral hepatitis.

Foods are occasionally involved in the spread of other viruses such as rotaviruses, tick-borne encephalitis virus, astroviruses, coronaviruses and enteroviruses (1,3). However, this publication will only consider noroviruses and HAV.

The Virus

Viruses are unique in nature. They are the smallest of all replicating organism and are characterised by their ability to pass through filters that retain even the smallest bacteria. They consist solely of a small segment of nucleic acid encased in a protein shell (4).

Noroviruses are human enteric calciviruses and are able to replicate only in the gastrointestinal tract of humans (4). Hepatitis A viruses belong to the family *Picornaviridae* (5).

Food Poisoning from Noroviruses

Noroviruses have accounted for 85% of non-bacterial gastrointestinal outbreaks in Europe between 1995-2000 (6). The provisional number of laboratory reports of norovirus infection in England and Wales for 2006 was 4,446 (Source: CDSC). Foodborne gastroenteritis often referred to as 'gastric or stomach flu', caused by noroviruses usually has a rapid (commonly 'explosive') onset of vomiting, abdominal pain and non-bloody diarrhoea, fever, nausea, chills, weakness, myalgia, and headache, accompanied in specific cases by a self-limiting course of illness (1,4) after an incubation time of about 24 h (range 15-48 h, depending on the agent; and on dose). Symptoms typically last about 24 h (range 12-60 h). Attack rates (the proportion of victims amongst those eating an implicated food) are usually high (often 50%). Virus is shed in vomit and in faeces for perhaps a week after onset (1). Numbers of virus particles in faecal material or vomit (which is often 'projectile' and uncontrollable) can be very high, so food and environment can easily become contaminated by an infected food handler, especially in catering (4). There is also frequent secondary spread from those infected to family and other contacts, so that the final number of people involved in an outbreak can be very large (numbers of cases occasionally reaching thousands) (2,4,6,7,8).

Viral gastroenteritis can occur in all age groups but it is more common in adults than in children.

Food Poisoning from Hepatitis A

Viral hepatitis transmitted by the enteric (faecal-oral) route (1) can be quite mild or even symptomless in children, where it is most common. The severity of the illness is greater in adults and increases with age (2,5).

The incubation time is normally about 4 weeks (15-50 days, median near 28 days) (1,5), and symptoms involve the gradual development of anorexia, malaise, pyrexia and vomiting, followed later by jaundice (1,9). Recovery is usually within a few weeks, but may take several months. Death has occurred (particularly amongst the elderly), but the fatality rate is low <1% (2,5,7,9).

Viral infective dose

Viruses need a 'host' in order to multiply - they are not capable of growing (or producing toxin) in foods. However, they may be extremely infectious, such that only a few particles (perhaps fewer than 10) need to be ingested to cause illness (4,5,7).

Foods involved

The main food type involved in reported incidents of illness involving foodborne viruses in the UK - whether hepatitis A or gastroenteritis - is molluscan shellfish, particularly raw oysters, as well as mussels and cockles. These shellfish are filter feeders that inhabit shallow coastal/estuarine waters, commonly near sewage outlets. The molluscs extract particles, including bacteria and viruses, from the large quantities of water passing over their gills. Depuration procedures that may be effective in eliminating bacterial pathogens cannot be relied upon to eliminate viruses, so the consumption of raw or inadequately cooked molluscs can lead to viral disease. The heat treatment of cockles to an internal temperature of 85-90 °C maintained for 1.5 min has been recommended for the destruction of HAV (1), and appears to have been effective in preventing outbreaks from norovirus in cooked molluscs (2,6,9).

The other group of foods that are associated with foodborne viral disease comprises foods that are handled or otherwise contaminated by infected food handlers, and that are not subsequently cooked, particularly in catering. Salads are especially vulnerable, as well as ice, prepared fruits like melons, soft fruits, dessert dishes, cakes, sandwiches, green onions, savoury snacks, soups etc., that may be heavily garnished and/or handled during preparation by hands that may be carrying viruses (2,4,6,10).

Sources of Foodborne Viruses

The original source of all foodborne viruses is the human intestine - the contamination of foods with either noroviruses or HAV is ultimately of human faecal origin (1,8,9,11).

Viruses can be carried in sewage-contaminated water, or on fresh produce such as vegetables harvested from soils treated with sewage sludge or polluted water. Faecal or vomit contamination of foods can also take place at the time of preparation or serving by infected food handlers (6,8,9,11).

Noroviruses can be excreted shortly before onset of symptoms, can be present in large numbers in stools and vomit during apparent illness, and decrease rapidly after recovery (usually within 24 h). HAV can be present in stools for up to 2 weeks before the onset of symptoms, but can be absent by about a week after the onset of jaundice (1,5,11).

Human 'carriers' of noroviruses or HAV have not been conclusively demonstrated (11).

Survival Characteristics of Foodborne Viruses

Because of the difficulties in studying foodborne viruses, especially noroviruses, very little is known about their survival characteristics in foods. However, norovirus has been shown to remain infective after exposure to pH 2.7 for 3 h, and has remained intact even after a 30 min exposure to pH 2 at body temperature (7). The virus may be inactivated by temperature treatments at 63 °C for 30 min, 70 °C for 2 min and within seconds at 100 °C (4). Also, like most viruses, they are resistant to refrigeration and freezing (1,8).

HAV is destroyed by heating at temperatures above 85 °C; it is resistant to acidic conditions (pH 3), as well as refrigerated and frozen storage (1,2,7,9,10).

Summary of Control

The lack of knowledge on these organisms means that the only practical means of control are adequate cooking of foods, prevention of cross-contamination and practising good personal and food hygiene.

Norovirus-infected personnel should be excluded from handling food for at least 2 days after recovery from viral gastroenteritis. The exclusion of infected food handlers is difficult in the case of HAV infection because the shedding of HAV particles in stools may have peaked well before jaundice symptoms become evident. However, it has been recommended that food handlers should be excluded for 7 days after the onset of jaundice in viral hepatitis (12).

PROTOZOA

Protozoa are single-celled organisms. They have compartmentalised organelles and the infectious stages are environmentally resistant. They are obligate intracellular organism and do not multiply in the environment. Thus studies of food matrices and determination of the safety of some products is challenging. In addition, protozoa cannot be cultured or cryopreserved in the way that bacteria can. Furthermore, cysts and oocysts may easily be mistaken, under microscopic examination, for other bodies, such as yeasts or mould spores. For these and other reasons, like viruses, protozoa are probably under-recognised as causes of foodborne and waterborne illness. Nevertheless, these parasites are becoming increasingly identified as a cause of gastroenteritis, especially in the immunocompromised, and as a cause of 'travellers' diarrhoea'.

Protozoa can be blood borne or transmitted via a vector, such as a mosquito. Others are acquired via water. Some cysts and oocysts may be present in potable water supplies from time to time, and present a problem for those in the food industry who use mains water in the preparation of foods that do not receive further heat treatment. The increased recognition of problems of contamination of water supplies with parasites, and consequent 'boil notices', led to the need to consider additional protection measures (for the purposes of 'due diligence') where mains water is used in food (and beverage) manufacture.

CRYPTOSPORIDIUM

The Organism

Cryptosporidium spp belong to the family Cryptosporidiidae. They were considered coccidian protozoan parasites. The most important species of

Cryptosporidium for humans are *C. hominis* and *C. parvum*. *C. hominis* almost exclusively infects humans, whereas *C. parvum* infects humans and livestock. Its usual host is food animals, especially lambs and calves, where it grows in the intestinal lining and is passed, as oocysts (the spore-like transmissible stage, size *ca* 4-6 microns) in the faeces. The life cycle (sexual and asexual) is completed in the gut of a single host (13,14,15,16,17).

Human Cryptosporidiosis

This is often a zoonotic disease, but direct person-to-person spread is common, especially amongst children. In the UK, infection is most common in children under six years old. Epidemiological evidence suggests that *Cryptosporidium* may also be transmitted via contaminated water, milk or food (13,14,15,17).

Incubation time

This depends on the size of the infective dose, but is typically 5-7 days (range 2-14 days).

Symptoms

Typically, symptoms comprise foul smelling, watery or mucoid diarrhoea, which may be accompanied by abdominal pain, vomiting, malabsorption, fever and loss of appetite and weight. Rarely, cryptosporidiosis may involve extra-intestinal organs including the gall bladder, lungs, eyes and vagina.

Mortality

In otherwise healthy individuals, symptoms commonly persist for 2 to 3 weeks and are self-limiting. However, hospitalisation is sometimes required and the illness may be life-threatening to the severely immunocompromised, such as AIDS patients. Cryptosporidiosis has been associated with increased child mortality in developing countries (15).

Infective dose

The infective dose for humans is not known but is thought to be as low as 10 oocysts, or even lower. (Infectivity studies with lambs have indicated a probable minimum infective dose of one oocyst.) (14,16)

Incidence of *Cryptosporidium* Infection

C. parvum is responsible for more infections in Europe and Kuwait and *C. hominis* is responsible for most human infections in the rest of the world.

The reported incidence of cryptosporidiosis appears to be on the increase both in the UK and North America. In 2006, there were about 3,500 reports of *Cryptosporidium* infection to the CDSC in England and Wales. It is not known how many of these can be attributed to water- or foodborne infection (18).

Sources

Humans

Person-to-person transmission is now considered to be common, primarily via the faecal-oral route. Oocysts may be excreted - initially in very large numbers - for from two weeks to two months after symptoms of infection cease. Repeated exposure may lead to immunity and asymptomatic infection; little is known about asymptomatic carriage but it is thought that asymptomatic cases may excrete low numbers of oocysts. Control is therefore reliant on scrupulous personal hygiene (14,17).

Animals and environment

Cryptosporidium spp have been reported in more than 40 species of mammals, birds, reptiles and fish. *C. parvum* is not host-specific and may be transmitted to man from a wide range of animals. The parasite develops in the gut and oocysts may be passed in extremely large numbers in the faeces (up to 10^6 daily), in a fully infective form. It is particularly common in young calves and lambs; an association has, therefore, been made between children bottle-feeding lambs and other farm-visit experiences. Pets appear to play a minor role in the transmission of the disease (14).

The disposal of animal excreta (or even human sewage) on farm land, e.g. through the practice of muck and slurry spreading, can lead to the contamination of water sources and supplies, as well as causing the direct contamination of food crops. Unusually heavy rainfall may contribute to the levels of oocysts in water supplies and exacerbate the above effect.

Food

Largely because cryptosporidia cannot be cultured in the laboratory, food has only rarely been linked directly to incidents of cryptosporidiosis. However, raw sausages, offal and raw milk have been suggested as possible vehicles (17). Fresh produce and shellfish that have been in contact with contaminated water are also possible vehicles of the organism. This is probably an important mode of transmission in travellers' diarrhoea (13,14,17).

However, it is generally considered that food is unlikely to be a major source of infection in comparison with animal-human (zoonotic), waterborne or person-to-person transmission.

Water

Oocysts are environmentally resistant and retain their infectious potential for considerable time in moist environments; they have been reported to survive for up to a year in seawater. Current water treatment systems cannot guarantee the complete absence of oocysts in mains water at all times (14,15).

In Britain, outbreaks of disease considered to be associated with inadequately filtered or contaminated drinking water have occurred in several parts of the country, some outbreaks involving hundreds of confirmed cases. A recent very large outbreak in Milwaukee, in the USA, affected an estimated 403,000 people, when the city water supply became contaminated with this parasite (19).

Air may also occasionally act as a vehicle for transmission of *Cryptosporidium*. The inhalation of oocysts held in aerosols in animal processing plants and during certain farming practices (such as muck spreading) may contribute to the occurrence of cryptosporidiosis.

Survival Characteristics of the Organism in Foods

Temperature

Oocysts can survive at refrigeration temperatures, but are killed by freezing (below -20 °C) (14).

Heat resistance

Oocysts are not heat-resistant, being sensitive to holding for 20 min or more at temperatures above 45 °C (14); they are readily destroyed by pasteurisation or heat treatments equivalent to 5-10 min at 65-85 °C (15). In the event of a 'boil notice', water need only be raised to boiling point and allowed to cool; a prolonged holding time is not required.

The organism is sensitive to desiccation, requiring moisture for survival. Air drying for a few hours at room temperatures should ensure complete destruction of any oocysts present. Oocysts do not survive freeze-drying (14).

pH, water activity and atmosphere

Little is known about the effects of pH, a_w and gas atmosphere on the survival of *Cryptosporidium* oocysts. However, extremes in pH appear to have a significant effect on survival.

Summary of Control of *Cryptosporidium*

As indicated, these parasites are not able to grow in food or water, requiring an animal or human host. However, the oocysts may remain viable outside the host for long periods under cool, moist conditions. They are also resistant to most disinfectants, such as chlorine; prolonged exposure to between 8,000 and 16,000 ppm chlorine may be necessary to kill oocysts. Filtration is necessary for their removal from water - 1 micron pore size or slow sand filtration with flocculation is recommended. Alternatively, an appropriate heat treatment, or treatment with 10 vol. hydrogen peroxide, chlorine dioxide, or ozone (e.g. 2 ppm for 10 min) may represent options, but each has its practical limitations. irradiation appears to require exposures greater than those achievable by conventional treatment (14,15).

GIARDIA

Like *Cryptosporidium*, *Giardia* causes gastroenteritis in humans; in fact it is now considered to be one of the leading protozoal causes of gastroenteritis worldwide.

The Organism

The species that is known to be pathogenic for man and animals is *G. lamblia* (also known as *G. intestinalis* or *G. duodenalis*). *Giardia* spp are binucleate flagellate protozoa, with both trophozoite and cyst stages. This organism is capable of infecting more than one host species. The importance of sheep and cows as sources of *Giardia* in the UK is not known, but it has been demonstrated in these animals, especially in calves and lambs. Cats and dogs may also be infected with *Giardia*, although the significance of this for humans is uncertain.

Giardiasis

Infection with *Giardia* is generally associated with poor hygienic conditions, including poor water quality (17).

Infection results from the ingestion of cysts, which may remain viable in cool, moist environments for long periods outside the host (13,16,18,20).

Incubation time

Symptoms occur after 1-4 weeks and generally last for about 2 weeks (range 5 days to several months if untreated), during which there may be shedding of cysts (cysts appear in stools after 3-4 weeks) and trophozoites in the stools.

Symptoms

Symptoms of illness include diarrhoea, malabsorption, abdominal pain, flatulence, loss of appetite and sometimes fever. Hospitalisation is occasionally required. Infection may also, commonly, be asymptomatic.

Mortality

Mortality has been reported only twice.

Infective dose

The infective dose is 10 or fewer cysts. Trophozoites are virtually non-infective owing to their susceptibility to gastric acid (13,16,18).

Sources

Three routes of transmission have been identified: food, water and person-to-person transmission. Only a small number of outbreaks - none of which has been identified as being foodborne - have been reported in the UK. These waterborne outbreaks were caused by post-treatment contamination of the water with sewage or faeces (18,20). Likewise, in the UK, it is not yet clear whether potable water is an important vehicle for the transmission of *Giardia* (18,21,22).

Child-care facilities, where poor personal hygiene is coupled with close physical contact between children, are a major site of person-to-person transmission.

In the USA, reported foodborne transmission - generally involving inadequate personal hygiene of exposed/infected food handlers - has implicated noodle- and fruit-salads, canned salmon and taco ingredients (17,23).

In spite of the fact that shellfish could concentrate *Giardia* cysts, to date, molluscs have not been involved in foodborne outbreaks. They could be a potential source of foodborne outbreaks though, especially when eaten raw (20).

Incidence of *Giardia* Infection

Outbreaks of giardiasis have been reported most often in the USA, where it is the most commonly recognised cause of waterborne outbreaks. It is estimated that 15% of the U.S population is infected with this organism (23).

In the UK, *Giardia* is rarely associated with large outbreaks, usually being reported from numerous apparently unconnected family or group cases. The incidence rose from 4,613 cases in 1983 to about 7,000 cases in

1992, but has since declined to just under 3,000 cases in 2006 (Source: CDSC).

Growth and Survival Characteristics of the Organism in Foods

Giardia cysts are easily killed by boiling, but they can survive freezing for a few days. They are, however, resistant to low pH values - above about pH 3. The cysts are susceptible to inactivation by ozone and halogens; however, the concentrations of chlorine used for drinking water may not inactivate *Giardia* cysts. Inactivation by chlorine requires prolonged contact time and filtration.

OTHER PROTOZOA

This section deals with other protozoa that may be foodborne or waterborne and have health implications.

Entamoeba histolytica

Entamoeba histolytica (cyst size 10-20 micron), is primarily a human parasite; dogs, cats and other mammals have been reported to be infected only rarely. The organism is anaerobic requiring glucose or galactose as its main respiratory substrate (23,24).

Incubation period is 2-4 weeks. Initially, infection causes amoebic dysentery (bloody or mucoid diarrhoea). Later, symptoms consist of abdominal pain, fever, severe diarrhoea, vomiting and lumbago; it may resemble shigellosis. Pregnant or nursing women may experience fulminating amebiasis with ulceration of the colon and subsequent toxicity. Asymptomatic infection is more common. There appear to be pathogenic and non pathogenic 'variants' of *E. histolytica*. These are morphologically indistinguishable, but it is now thought that they may represent different species (24).

If trophozoites (12-60 μm) of pathogenic entamoeba enter the bloodstream, they may enter the liver (or, rarely, other organs), with serious consequences (13,23,24).

Transmission of the cyst is primarily through the person-to-person faecal-oral route. However, food or water can become contaminated with cysts by food handlers, where human excrement is used as fertiliser for crops, or where foods that are not further cooked are exposed to sewage-polluted water (encysted forms can survive as long as 3 months in sewage sludge) (13,23).

The cysts are less resistant to chlorine than *Cryptosporidium*, and are also sensitive to heat and freezing. As with *Giardia*, trophozoites are virtually non-infective (13).

Toxoplasma

The coccidian parasite, *Toxoplasma*, contains one species, *T. gondii*. Its definitive host is the cat family. Oocysts may be excreted by cats in large numbers and may survive for over a year under cool, moist conditions. Initially, they are not infective, but become infective after 24-48 h exposure to air. Human infection can follow the ingestion of as few as 100 oocysts (incubation period in adults is 6-10 days while in infants it is congenital); waterborne outbreaks have also been associated with *Toxoplasma*. Other animals, such as livestock (sheep, goat, chicken and pork), can become infected through exposure to pasture or feed that has been contaminated by cat faeces. Human infection can then follow ingestion of raw or undercooked meat from such animals (13,21,23,25).

It is thought that foodborne infection in man is uncommon. The parasite is sensitive to heat (>61 °C for 3.6min) and freezing (-13 °C) (20). However, handling raw meat and the consumption of raw meat dishes may represent a hazard, particularly to pregnant women as it can be passed on from mother to child, *in utero*. (20,23,25).

Infection in humans can be asymptomatic or can occur as a mild 'flu-like' illness with rash, headache, muscle aches and pain and swelling of the lymph nodes However, congenital infections or infection in the immunocompromised can be serious. Latent, asymptomatic infections can become active and life-threatening upon loss of immunocompetence (due to medications or disease, especially AIDS) (23).

Balantidium coli

Balantidium coli are ciliated protozoa (70 μm diameter) associated with a variety of hosts, especially pigs and non-human primates. Infection in man follows ingestion of cysts; symptoms include diarrhea or dysentery, tenesmus, nausea, vomiting, anorexia and headache. Insomnia, muscular weakness and weight loss has also been reported. It can also cause ulcerative colitis. Ulcers differ from those caused by *Entamoeba histolytica* in that the epithelial surface is damaged, but with more superficial lesions compared to those caused by amoebae. The infection can be transmitted from person to person or via food (13,20,24).

Sarcocystis

Sarcocystis are predominantly coccidian parasites of cattle and pigs belonging to the family Sarcocystidae. Infection in man is acquired by the consumption of raw or undercooked meat (especially beef or pork) that contains the cysts. Studies have shown that symptoms occur within 3-48 h (depending on the species) and consist of abdominal pain, fatigue, dizziness, nausea, vomiting, diarrhoea and malaise. Muscle ache is another symptom associated with *Sarcocystis* infection (13,23,26).

Microsporidia

Microsporidia are obligate intracellular spore-forming parasites found in mammals, fish, crustaceans and insects. This parasite is transmitted by the faecal-oral route by means of small spores (*ca* 1 micron); the source is not known but the faecal excretion route provides the potential for water- or foodborne transmission. Infection occurs mainly in AIDS patients with enteritis being the most common clinical manifestation of microsporidiosis. Diarrhoea is typically chronic and intermittent, loose to watery, and non-bloody. Anorexia, weight loss and dehydration are common, and abdominal cramping, nausea and vomiting may occur. Other clinical features include keratoconjunctivitis, mycositis, nephritis, hepatitis, sinusitis and pneumonia. (20).

Cyclospora

Cyclospora spp are coccidian protozoa that belong to the family Eimeriidae. These parasites are common in a variety of mammals and cause disease when infectious oocysts are ingested; although the definitive host for the species infecting man is not yet known. The route of transmission is faecal-oral. Infection in otherwise healthy individuals can cause diarrhoea, malaise, lack of energy and appetite, and weight loss lasting for 4-8 weeks. Incubation period ranges between 2 and 11 days. Infection occurs most commonly in those travelling to developing countries. Infections have been reported from such places as Nepal, India, Pakistan, Indonesia, Southeast Asia, Papua New Guinea, Middle East, North Africa, the United Kingdom, the Carribean, the United States, and Central and South America, although the true prevalence in any population is unknown. There is some evidence for waterborne transmission but this is controversial. There have been a number of well documented outbreaks in the USA associated with imported fresh produce like berries, basil, lettuce and snow peas. (20,23,26).

Bibliography

References

1. Cliver D.O. Viruses, in *Guide to Foodborne Pathogens*. Eds Labbe R.G., Garcia S. New York. John Wiley & Sons, Inc. 2001, 257-69.

2. Richmond M. Viral foodborne disease. Appendix 3, Committee on the Microbiological Safety of Food, The Microbiological Safety of Food, Part 2. Report of the Committee. London. *HMSO*, 1991, 195-9.

3. Cliver D.O. Foodborne viruses, in *Food Microbiology: Fundamentals and Frontiers*. Eds Doyle M.P., Beuchat L.R., Montville T.J. Washington DC. ASM Press, 2001, 501-12.

4. Duizer E., Koopmans M. Tracking emerging pathogens: The case of noroviruses, in *Emerging Foodborne Pathogens*. Eds Motarjemi Y., Adams M. Cambridge. Woodhead Publishing Ltd, 2006, 77-110.

5. Cook N., Rzezutka A. Hepatitis viruses, in *Emerging Foodborne Pathogens*. Eds Motarjemi Y., Adams M. Cambridge. Woodhead Publishing Ltd, 2006, 282-308.

6. Goyal S. Viruses in foods. New York. Springer, 2006.

7. Cliver D.O., Matsui S.M., Casteel M. Infections with viruses and prions, in *Foodborne Infections and Intoxications*. Eds Riemann H.P., Cliver D.O. London. Academic Press, 2006, 367-448.

8. Appleton H. Norwalk virus and the small round viruses causing foodborne gastroenteritis, in *Foodborne Disease Handbook, Vol. 2. Diseases Caused by Viruses, Parasites, and Fungi*. Eds Hui Y.H., Gorham J.R., Murrell K.D., Cliver D.O. New York. Marcel Dekker, 1994, 57-79.

9. Cromeans T., Nainan O.V., Fields H.A., Favorov M.O., Margolis H.S. Hepatitis A and E viruses, in *Foodborne Disease Handbook, Vol. 2. Diseases Caused by Viruses, Parasites, and Fungi*. Eds Hui Y.H., Gorham J.R., Murrell K.D., Cliver D.O. New York. Marcel Dekker, 1994, 1-56.

10. Varnam A.H., Evans M.G. Foodborne viral infections, in *Foodborne Pathogens: An Illustrated Text*. Eds Varnam A.H., Evans M.G. London. Wolfe Publishing Ltd, 1991, 363-72.

11. Cliver D.O. Epidemiology of viral foodborne disease. *Journal of Food Protection I*, 1994, 57 (3), 263-6.

12. Department of Health, Management of Outbreaks of Foodborne Illness. Heywood. Department of Health, 1994, 135.

13. Varnam A.H., Evans M.G. Protozoa, in *Foodborne Pathogens: An Illustrated Text*. Eds Varnam A.H., Evans M.G. London. Wolfe Publishing Ltd, 1991, 373-86.

14. Casemore D.P. Epidemiological aspects of human cryptosporidiosis. *Epidemiology and Infection*, 1990, 104 (1), 1-28.

15. Xiao L., Cama V. *Cryptosporidium* and Cryptosporidiosis, in *Foodborne Parasites*. Ed. Ortega Y.R. New York. Springer, 2006, 57-86.

16. Eley A.R. Viruses and protozoa, in *Microbial Food Poisoning*. Ed. Eley A.R. London. Chapman & Hall, 1996, 95-108.

17. Casemore D.P. Foodborne illness - foodborne protozoal infection. *Lancet*, 1990, 336 (8728), 1427-32.

18. Tully M.J. *Giardia intestinalis*: the organism and its epidemiology. *PHLS Microbiology Digest*, 1993, 10 (3), 129-32.

19. MacKenzie W.R., Hoxie N.J., Proctor M.E., Gradus M.S., Blair K.A., Peterson, D.E., Kazmierczak J.J., Addiss D.G., Fox K.R., Rose J.B., Davis J.P. A massive outbreak in Milwaukee of *Cryptosporidium* infection transmitted through the public water supply. *New England Journal of Medicine*, 1994, 331 (3), 161-7.

20. Ortega Y. Foodborne and waterborne protozoan parasites, in *Foodborne Pathogens: Microbiology and Molecular Biology*. Eds Fratamico P.M., Bhunia A.K., Smith J.L. Wymondham. Caister Academic Press, 2005, 145-61.

21. Smith H.V., Robertson L.J., Campbell A.T., Girdwood R.W.A. *Giardia* and giardiasis: what's in a name? *Microbiology Europe*, 1995, 3 (1), 22-9.

22. Flanagan P.A. *Giardia* - diagnosis, clinical course and epidemiology. A review. *Epidemiology and Infection*, 1992, 109 (1), 1-23.

23. Jay J.M., Loessner M.J., Golden D.A. Foodborne animal parasites, in *Modern Food Microbiology*. Eds Jay J.M., Loessner M.J., Golden D.A. New York. Springer Science, 2005, 679-708.

24. Ortega Y.R. Amoeba and ciliates, in *Foodborne parasites*. Ed. Ortega Y.R. New York. Springer, 2006, 1-14.

25. Ortega Y.R. Toxoplasmosis, in *Foodborne parasites*. Ed. Ortega Y.R. New York. Springer, 2006, 109-133.

26. Cama V. Coccidian parasites, in *Foodborne parasites*. Ed. Ortega Y.R. New York. Springer, 2006, 33-55.

Further reading

Mattison K., Karthikeyan K., Abebe M., Malik N., Sattar S.A., Farber J.M., Bidawid S. Survival of calicivirus in foods and on surfaces: experiments with feline calicivirus as a surrogate for norovirus. *Journal of Food Protection*, 2007, 70 (2), 500-3.

Croci L., Losio M.N., Suffredini E., Pavoni E., di Pasquale S., Fallacara F., Arcangeli G. Assessment of human enteric viruses in shellfish from the northern Adriatic Sea. Various authors. WHO surveillance programme for control of foodborne infections and intoxications in Europe - January 2007. *WHO Surveillance Newsletter*, 2007, (January), (88-89), 1-5.

Ortega Y.R., Torres M.P., Van Exel S., Moss L., Cama V. Efficacy of a sanitizer and disinfectants to inactivate Encephalitozoon intestinalis spores. *Journal of Food Protection,* 2007, 70 (3), 681-4.

Ortega Y.R., Liao J. Microwave inactivation of *Cyclospora cayetanensis* sporulation and viability of *Cryptosporidium parvum* oocysts. *Journal of Food Protection*, 2006, 69 (8), 1957-60.

Mena K.D. Risk assessment of parasites in food, in *Foodborne parasites*. Ed. Ortega Y.R. New York. Springer, 2006, 275-84.

Erickson M.C., Ortega Y.R. Inactivation of protozoan parasites in food, water, and environmental systems. *Journal of Food Protection*, 2006, 69 (11), 2786-808.

Sulaiman I.M., Cama V. The biology of *Giardia* parasites. *Foodborne parasites*. Ed. Ortega Y.R. New York. Springer, 2006, 15-32.

Goyal S. Viruses in foods. New York. Springer, 2006.

Ortega Y.R. Foodborne parasites. New York. Springer, 2006.

Cannon J.L., Papafragkou E., Park G.W., Osborne J., Jaykus L.-A., Vinje J. Surrogates for the study of norovirus stability and inactivation in the environment: a comparison of murine norovirus and feline calicivirus. *Journal of Food Protection*, 2006, 69 (11), 2761-5.

D'Souza D.H., Sair A., Williams K., Papafragkou E., Jean J., Moore C., Jaykus L. Persistence of caliciviruses on environmental surfaces and their transfer to food. *International Journal of Food Microbiology*, 2006, 108 (1), 84-91.

Chancellor D.D., Tyagi S., Bazaco M.C., Bacvinskas S., Chancellor M.B., Dato V.M., de Miguel F. Green onions: potential mechanism for hepatitis A contamination. *Journal of Food Protection*, 2006, 69 (6), 1468-72.

Friedman D.S., Heisey-Grove D., Argyros F., Berl E., Nsubuga J., Stiles T., Fontana J., Beard R.S., Monroe S., McGrath M.E., Sutherby H., Dicker R.C., DeMaria A., Matyas B.T. An outbreak of norovirus gastroenteritis associated with wedding cakes. *Epidemiology and Infection*, 2005, 133 (6), 1057-63.

Korsager B., Hede S., Boggild H., Bottiger B., Molbak K. Two outbreaks of *Norovirus* infections associated with the consumption of imported frozen raspberries, Denmark, May-June 2005. *Eurosurveillance Weekly*, 2005, 10 (6).

Shaw I. Viruses. *Is it safe to eat? Enjoy eating and minimize food risks*. Ed. Shaw I. Berlin. Springer Verlag, 2005, 79-96.

Montville T.J., Matthews K.R. Viruses and prions. *Food microbiology: an introduction*. Eds Montville T.J, Matthews K.R. Washington DC. ASM Press, 2005, 289-97.

Richards G.P. Foodborne and waterborne enteric viruses. *Foodborne pathogens: microbiology and molecular biology*. Eds Fratamico P.M., Bhunia A.K., Smith J.L. Wymondham. Caister Academic Press, 2005, 121-43.

Jay J.M., Loessner M.J., Golden D.A.Viruses and some other proven and suspected foodborne hazards. *Modern food microbiology*. Jay J.M., Loessner M.J., Golden D.A. New York. Springer Science, 2005, 727-45.

Dawson D. Foodborne protozoan parasites. *International Journal of Food Microbiology*, 2005, 103 (2), 207-27.

Williams K.E., Jaykus L.-A. Norwalk-like viruses and their significance to foodborne disease burden. *Journal of the Association of Food and Drug Officials*, 2002, 66(4), 28-42.

Seymour I.J., Appleton H. A review: foodborne viruses and fresh produce. *Journal of Applied Microbiology*, 2001, 91(5), 759-73.

Taylor M.A. Protozoa. *The microbiological safety and quality of food, volume 2.* Eds Lund B.M., Baird-Parker T.C., Gould G.W. Gaithersburg. Aspen Publishers, 2000, 1420-56.

Caul E.O. Foodborne viruses. *The microbiological safety and quality of food, volume 2.* Eds Lund, B.M., Baird-Parker, T.C., Gould, G.W. Gaithersburg. Aspen Publishers, 2000, 1457-89.

Lees D. Viruses and bivalve shellfish. *International Journal of Food Microbiology,* 2000, 59 (1-2), 81-116.

Hui Y.H., Sattar S.A., Murrell K.D., Nip W.K., Stanfield P.S. *Foodborne disease handbook, volume 2: viruses, parasites, pathogens and HACCP, 2nd edition.* New York. Marcel Dekker, 2000, 515.

Soave R., Herwaldt B.L., Relman D.A. *Cyclospora. Infectious Disease Clinics of North America,* 1998, 12 (1), 1-12.

Gelinas P. Protozoa. *Handbook of Foodborne Microbial Pathogens.* Ed. Gelinas P. Morin Heights. Polyscience Publications, 1997, 121-36.

Marshall M.M., Naumovitz D., Ortega Y., Sterling C.R. Waterborne protozoan pathogens. *Clinical Microbiology Reviews,* 1997, 10 (1), 67-85.

Cliver D.O. Virus transmission via food. *Food Technology,* 1997, 51 (4), 71-8.

Anon. Cyclosporiasis in north America associated with fruit and vegetables. *CDR Weekly.* 1997, 7 (32), 283,286.

Various authors. Outbreak of cyclosporiasis - northern Virginia-Washington, DC-Baltimore, Maryland, Metropolitan Area, 1997. *Morbidity and Mortality Weekly Report,* 1997, 46 (30), 689-91.

Laberge I., Griffiths M.W., Griffiths M.W. Prevalence, detection and control of *Cryptosporidium parvum* in food. *International Journal of Food Microbiology,* 1996, 32 (1-2), 1-26.

International Commission on Microbiological Specifications for Foods. Viruses. *Microorganisms in Foods, Vol. 5. Microbiological Specifications of Food Pathogens.* London. Blackie, 1996, 440-57.

Fayer R. Foodborne and waterborne zoonotic protozoa. *Foodborne Disease Handbook, Diseases Caused by Viruses, Parasites and Fungi.* Eds Hui Y.H, Gorham J.R, Murrell K.D, Cliver D.O. New York. Marcel Dekker, 1994, 331-62.

Smith J.L. *Cryptosporidium* and *Giardia* as agents of foodborne disease. *Journal of Food Protection,* 1993, 56 (5), 451-61.

FOOD-SPOILAGE BACTERIA

INTRODUCTION

Microbial Food Spoilage – a Brief Overview

Food spoilage can be defined as any change in the characteristics of a food that renders it unacceptable. The cause of such a change may be a physical process, such as moisture loss, a chemical process, such as rancidity, or a biological process, such as insect infestation or microbial action. In some cases, spoilage may be the result of a combination of two or more of these. However, it can be said that the most common cause of food spoilage is microbiological activity.

The economic cost of microbial food spoilage is difficult to estimate, but it is certainly large. Despite the benefits of modern preservation and storage technology, considerable quantities of food have to be discarded each year as a result of the actions of micro-organisms. Even so, spoilage receives relatively little attention in comparison with food safety, and the processes and interactions involved are often not well understood. Microbial food spoilage is also important in many foods as a warning of the possible growth of pathogens, and, although the difference between spoiled and unsafe foods may be clear to microbiologists, it is worth remembering that the consumer does not necessarily make such a distinction. In any case, products that spoil before the end of their shelf life may have to be withdrawn from sale, since the consumer will not buy spoiled food. A scan of recall notices in the press will reveal that a high proportion of the recalls due to microbiological problems are concerned with spoilage, not food pathogens.

Microbial food spoilage is usually due to the presence, survival and eventual growth of bacteria, yeasts, or moulds in a food product. Spoilage may then be apparent as visible growth, discoloration, gas production, or changes in texture, but is more usually due to 'off'-odours or taints from the by-products of microbial metabolism.

There are few foods that are inherently resistant to microbiological spoilage. Frozen foods will not support microbial growth, and neither will products with very low water activities, although adequate storage conditions are essential to maintain low moisture levels. Consumers now favour lightly preserved and minimally processed foods, and microbial spoilage is therefore a potential problem with most products. Whether spoilage will occur, and how rapidly, are essentially questions of microbial ecology and are determined by a combination of environmental and other factors. In fact, there are now more instances of spoilage in these types of foods, and raw ingredient quality and processor hygiene are now more important than ever.

Factors affecting the growth of micro-organisms in foods

Factors that control microbial growth in foods can be divided into four main categories:

Intrinsic factors

The inherent physical, chemical and biological properties of the food, such as pH, redox potential, water activity and presence of antimicrobial substances.

Extrinsic factors

The characteristics of the environment in which the food is maintained, such as the temperature, atmosphere and relative humidity.

Implicit factors

Properties of the microbial population present on a food, such as growth rates, and interactions between the different elements of the population. These interactions may be synergistic or antagonistic.

Processing factors

The physical or chemical operations involved in food processing, such as pasteurisation, washing, peeling or slicing.

It is these factors, and the interactions between them, that define the environment within a particular food product, and determine the nature of the resident microbial population and its potential to grow and eventually cause spoilage.

Spoilage of unprocessed foods

Unprocessed foods, such as meat, fish, raw milk or vegetables, usually support a varied, natural microflora. Many species are likely to be present, and some of these will be able to exploit the environment provided by the food more effectively than others. These organisms will eventually predominate, even though they may have been only a minor component of the initial population, and will normally become the major causes of spoilage. Since the initial population of unprocessed foods is naturally diverse, these principal spoilage species are always likely to be present, and, for this reason, the composition of the microbial population at the point of spoilage is quite predictable. The microbial populations that develop during the spoilage of many unprocessed foods have been extensively studied and reported. For example, large numbers of *Pseudomonas* spp. are almost always found in spoiled, chilled, fresh meats.

Sometimes, however, the major microbial causes of spoilage do not predominate. For example, it has been estimated that the main species of spoilage bacteria in chilled fresh fish make up only about 30% of the total population when spoilage becomes apparent. The predominant organisms in this case are not really involved in spoilage. This is a reflection of the particular nature of the biochemical changes that cause the fish to become unacceptable. The interactions between the spoilage microflora and the biochemical processes that occur during spoilage are not fully understood for many foods.

For some foods, it has been possible to identify specific spoilage organisms (SSOs) within the general spoilage population. For example *Photobacterium phosphoreum* has been reported to be primarily responsible for spoilage in certain fish products, although it may not predominate. For many foods, SSOs have not yet been found, but this is a potentially interesting area for future research, since it may be possible to target the SSO in a particular product as a means of extending shelf-life.

Microbial spoilage does not usually become apparent until the population reaches quite high levels (typically $>10^7$ per g for bacteria). This is because individual bacteria have only a very small potential to cause biochemical changes in foods, and the collective activity of a large population is required before those changes become detectable. However, there is evidence that this is not always the case. For example, relatively small populations (approximately 10^3 per g) of *Alicyclobacillus acidoterrestris* have been shown to be sufficient to produce a taint in fruit-based beverages. Very small starting populations of *Zygosaccharomyces* spp. can also grow to sufficient levels to produce gas and cause problems in beverages.

Spoilage of processed foods

Processed foods are likely to have been subjected to treatments designed to delay the onset of spoilage, such as pasteurisation, modified-atmosphere packaging or drying. In other words, the factors that affect the growth of micro-organisms have been manipulated. Very often this results in eventual spoilage that is less immediately predictable. The initial microbial population of a processed food may be reduced and potential spoilers may have been eliminated (e.g. by heating). Therefore, other organisms, not normally regarded as spoilers, may be able to predominate and cause spoilage. For example, heat-resistant lactobacilli may cause spoilage of vacuum-packed, cooked meats, and psychrotrophic bacilli may spoil pasteurised milk products.

Spoilage bacteria

Although spoilage of foods by yeasts and moulds is important, particularly in acid and low-moisture products, bacterial spoilage is potentially a more serious problem with highly perishable products, especially chilled foods. Bacterial spoilage may be caused by a diverse range of species, may give rise to varied spoilage symptoms, and is not always predictable in processed foods. Therefore, it is important to understand something of the physiology and ecology of potential spoilage bacteria. This knowledge provides a possible basis for the identification of potential spoilage problems in new products, and the investigation of problems with existing ones. This section of Micro-Facts is intended to provide concise summaries of the key characteristics of a range of important spoilage bacteria.

In some cases, very little has been published about the attributes of some of the bacterial species involved in spoilage. Non-pathogenic bacteria in foods have been much less extensively studied than those that cause food poisoning. Nevertheless, the following pages represent an attempt to compile the key points about each genus, as far as is currently possible.

We have updated the data for this edition, and supplemented this with new sections on spoilage by yeasts and moulds.

ACETIC ACID BACTERIA

The Organisms

Acetic acid bacteria are aerobic, Gram-negative, oxidase-negative, rod or ellipsoidal-shaped cells that can typically oxidise ethanol to acetic acid. They belong to the family Acetobacteraceae comprising 14 genera, of which *Acetobacter*, *Asaia*, *Acidomonas*, *Gluconoacetobacter*, *Gluconobacter* and *Kozakia*, have been associated with food. Some species of *Gluconacetobacter* (a recently described genus) were formerly classified as *Gluconobacter* or *Acetobacter* species.

Acetic acid bacteria have been defined as bacteria capable of acetification, i.e. the aerobic conversion of ethanol to acetic acid. *Acetobacter*, but not *Gluconobacter*, can cause over-oxidation of formed acetic acid, to produce carbon dioxide and water.

Key Species in Foods

Acetobacter spp and *Gluconobacter* spp are well known contaminants of alcoholic beverages, and *Gluconobacter* is often associated with soft drinks and juices. *Acetobacter* and *Gluconobacter* are also associated with fruit spoilage problems, such as pink disease in pineapples, and rots in apples and pears.

Acetic acid bacteria are used commercially in the vinegar production industry (usually *Acetobacter* spp).

More recently, the newly described *Gluconoactebacter sacchari* and *Asaia* sp. have been isolated from fruit-flavoured bottled water.

Sources

Environment

Acetic acid bacteria are widespread in nature and their natural habitats are sugared and alcoholised niches such as flowers and fruit. *Gluconobacter* have been isolated from the nectar bearing parts of flowers, as well as from bees and their hives. *Acetobacter* can also be found in garden soil and canal water. Although both *Acetobacter* and *Gluconobacter* have been isolated from fruits, *Acetobacter* are usually found when fermentation is more advanced because of their preference for alcohol-enriched environments.

Apart from flowers and fruit, *Asaia* have also been found in fermented glutinous rice, and *Gluconacetobacter* spp have been isolated from various plant materials such as sugar cane and coffee plants.

Food

Gluconobacter and *Acetobacter* are associated with alcoholic beverages such as beer and wine. *Gluconobacter* spp have been isolated from low pH, high sugar soft drinks and juices, as well as fruit. *Acetobacter* have been isolated from spoiled lactic acid fermented meat products.

Spoilage Characteristics

In soft drinks, particularly still drinks, spoilage by acetic acid bacteria classically causes turbidity in the product and a characteristic sour odour. In these products, spoilage by acetic acid bacteria is usually caused by *Gluconobacter* spp because of their preference for high-sugar environments. Sometimes off-flavours can be the only indication that spoilage has occurred and some strains can be present in high numbers but have no effect on the taste or odour of the product. Turbidity and off-flavours in fruit-flavoured bottled water caused by *Asaia* sp and *Gluconoacetobacter sacchari* have also been described.

In beer, wine, sake and, to a lesser extent, cider, acetic acid bacteria can cause vinegary off-flavours, turbidity, discoloration and ropiness (or slime) as a result of extracellular cellulose formation. The surface contamination they cause is often apparent as an oily or mouldy film. *Acetobacter* is of the most importance in these products because of its preference for alcohol-enriched niches and, in wines, *Acetobacter pasteurianus* can cause a vinegary or a mousy, sweet-sour taste.

A visual defect in stored vinegar, known as 'mother of vinegar', is caused by *Acetobacter* species. In lactic acid fermented meat commodities, overt odourous spoilage by *Acetobacter* strains has been recorded and the growth of these organisms can lead to inhibition of the lactic acid bacteria, resulting in potential food safety issues.

Growth/Survival Characteristics of the Organisms in Foods

Temperature

The optimum for growth of acetic acid bacteria is 25-30 °C. However, some *Acetobacter* strains are recorded as growing at 37 °C and above, and certain *Gluconobacter* strains may be able to grow at temperatures as low as 4 °C.

Acetic acid bacteria are not particularly heat-resistant, and are not reported as surviving mild pasteurisation processes.

pH

The optimum pH for growth of acetic acid bacteria is around pH 5-6, with a minimum pH for growth around range of 3-3.5, for most acetic acid bacteria. Many *Gluconobacter* strains can remain viable at pH 2.4. *Acetobacter acidophilus* can grow at pH 2.4 and some industrial strains of *Gluconobacter* can tolerate acetic acid concentrations up to 13.5%.

Water activity

Although growth of these organisms does not occur at low water activity, they may survive in concentrates and cause spoilage once the concentrate is diluted. The minimum a_w for growth is likely to be dependent on many factors including solute present. As previously noted, *Gluconobacter* spp. have a preference for high sugar niches, whereas *Acetobacter* favours alcohol-enriched environments.

Atmosphere

Acetic acid bacteria are obligate aerobes and are usually not a problem in highly carbonated beverages because of their sensitivity to elevated levels of dissolved carbon dioxide.

Preservatives and biocides

Strains of *Gluconobacter* and *Acetobacter* resistant to sorbate and benzoate have been reported, and acetic acid bacteria have the potential to cause the spoilage of products that rely solely upon these preservatives for their stability.

Control

The main means of control of the acetic acid bacteria is the use of mild heat treatments. The use of preservatives can prove effective, particularly in combination with other hurdles such as reduced water activity and titratable acidity. However, resistant strains can develop. Care should be taken in choosing raw materials of good microbiological quality and to ensure good hygiene in processing equipment when producing products that may be susceptible to acetic acid bacteria spoilage.

ACINETOBACTER

The Organism *Acinetobacter*

Acinetobacter spp are Gram-negative, non-motile, non-sporeforming rods. They can become spherical during the stationary phase of the growth curve. They are aerobes, with a strictly respiratory metabolism, and are oxidase-negative and catalase-positive. The genus *Acinetobacter* is quite heterogeneous, belonging to the *Moraxellaceae* family.

Key Species in Foods

Currently there are 17 genomic species, but the species that have been most commonly isolated from foods include *A. lwoffi*, *A. johnsonii* and *A. calcoaceticus*.

Sources

Environment

Acinetobacter spp are common and widely distributed in soil, water and sewage. They are also found as components of the normal skin microflora of humans and animals.

Foods

These organisms may be isolated from a wide variety of foods, particularly chilled fresh meat, fish, eggs, poultry, milk and vegetables. They have also been found to cause spoilage of soft drinks.

Spoilage Characteristics

Acinetobacter spp are an important part of the normal spoilage flora of fresh meat stored under aerobic conditions. They predominate under conditions of temperature abuse, but are less important at chill temperatures, where Pseudomonads predominate. They produce 'off'-odours as a result of amino acid metabolism and may also cause souring of ham. They are common spoilage organisms of poultry and shellfish and may cause 'colourless' rots in eggs.

Spoilage of soft drinks by *A. johnsonii* was characterised by white filamentous growth.

Growth/Survival Characteristics of the Organisms in Foods

Temperature

Some strains are psychrotrophic and growth at 4 °C may occur, but *Acinetobacter* spp do not compete well under these conditions. Strains grow well between 20 °C and 30 °C, the optimum temperature for most strains being 33-35 °C.

pH

It has been reported that growth is reduced on meats with a pH of <5.7, and the optimum pH for growth seems to be 5.5-6.0. In some cases *Acinetobacter* has been able to tolerate low pH; growth has been reported in soft drinks with pH 3.3.

Water activity

A minimum a_w value permitting growth of 0.96 has been reported for a strain identified as *Acinetobacter* spp, but a value as high as 0.99 has been reported for a meat-spoilage isolate. No *Acinetobacter* were isolated in a study on fish fillets treated with 10% brine.

Atmosphere

Acinetobacter spp are aerobic organisms and are not normally involved in the spoilage of vacuum-packed or similar products. Despite this fact they have been isolated in low numbers from chilled vacuum-packed meats.

Preservatives and biocides

There is very little published information on the effect of antimicrobials on *Acinetobacter* spp. They are known to withstand exposure to 0.7% hydrogen peroxide and 10 ppm ozone.

Control

Acinetobacter spp are normally present on the surface of fresh meat and poultry carcases, and may cause significant spoilage under aerobic conditions, particularly if the temperature is allowed to rise above 8 °C. Spoilage is therefore best controlled by careful temperature control. As they do not tolerate high levels of salt and are not resistant to heat, growth can be controlled by increasing the salt concentration and by pasteurisation.

Prevention of cross-contamination from the skin during processing, by good hygienic practices, is also important.

ALICYCLOBACILLUS

The first *Alicyclobacillus* species was isolated in 1982 and was initially classified as a *Bacillus*. It has since been reclassified into the new genus *Alicyclobacillus* partly because of the presence of ω–alicyclic fatty acid in their cell membrane.

The Organism *Alicyclobacillus*

Alicyclobacillus spp are Gram-positive, thermophilic and acidophilic sporeforming (terminal or subterminal with or without swollen sporangia), facultative anaerobic motile rods.

Key Species in Foods

Only one species, *A. acidoterrestris*, has so far been isolated from foods. Two other species, *A. acidocaldarius* and *A. cycloheptanicus*, have been identified, but these are thought to be mainly associated with hot springs.

Sources

Environment

Soil is the probable natural habitat of *A. acidoterrestris*.

Foods

The organism has been isolated from low pH (e.g. <3.8) canned/ hermetically sealed products such as fruit juices, juice-based beverages and tomato-based products. It has also been isolated from more diverse shelf-stable products, including spoilage of carbonated fruit drinks and iced tea. It is probably present on the surface of whole fruits as a result of soil contamination.

Water is also a possible source of contamination; in particular, water used for beverage manufacturing processes.

Spoilage Characteristics

Spoilage takes the form of an 'antiseptic', 'medicinal' or 'phenolic' taint in fruit beverages, particularly in apple-based, pasteurised, shelf-stable fruit drinks. The offensive taste or odour has been attributed to guaiacol, bromophenols and chlorophenols; these are potent tainting compounds. A slight sediment may also be present, but no gas is produced (flat-sour spoilage). Spoilage can result from very low initial levels of contamination, and taints can be produced by populations as low as 10^3 cfu/ml.

Growth/Survival Characteristics of the Organisms in Foods

Temperature

A. acidoterrestris is essentially a thermophile, but isolates seem able to grow over a very wide range of temperatures; growth at temperatures as low as 12 °C and as high as 80 °C has been reported, but growth is very slow at the extremes of this range. The optimum growth temperature seems to be 42-60 °C.

The spores of the organism are very resistant to elevated temperature and are able to survive pasteurisation processes applied to fruit juices. D-values at 90 °C of 16-23 min have been reported for a spoilage strain. Heat resistance is also much affected by the pH of the heating medium, and small increases in pH may produce significantly higher resistance.

pH

A. acidoterrestris is an acidophile and is able to grow within the pH range 2.0-6.0, with an optimum pH of 3.5-5.0.

Water activity

The organism does not seem to grow at reduced water activities. Inhibition by sugar concentrations of 18 ºBrix has been reported, and growth does not occur in fruit juice concentrates.

Atmosphere

A. acidoterrestris is aerobic, but has been reported to grow in sealed beverage containers.

Preservatives and biocides

A concentration of 6% ethanol inhibits growth, but preservatives added to juices, such as benzoic acid and sorbic acid have also been shown to prevent spoilage.

Control

It has been recommended that control requires the monitoring of raw materials for contamination by *A. acidoterrestris*, so that contaminated fruit can be excluded from vulnerable products. Effective washing of fruit surfaces has also been suggested as a control measure. In addition, storage of fruit juices in refrigerators and reducing head space will slower growth rate of Alicyclobacilli. Spoilage by this organism is rare, but can be a serious problem.

BACILLUS

The genus *Bacillus* is large and diverse, and includes a number of species important in foods. Several are pathogens and some are important in food spoilage.

The Organism *Bacillus*

They are Gram-positive, endospore-forming rods, and most are motile. They are aerobic or facultatively anaerobic, catalase-positive and oxidase-positive or -negative. They are able to carry out a wide range of metabolic processes, and the genus includes psychrotrophic and thermophilic species. Gas is not usually produced from glucose. Endospores are resistant to environmental factors and may remain viable for many years.

Key Species in Foods

Some of the species known to cause foodborne disease are also important in food spoilage, particularly *B. cereus*, *B. subtilis* and *B. licheniformis*. Other species, such as *Paenibacillus polymyxa* (formerly *Bacillus polymyxa*) and *B. circulans*, are also isolated from foods. *B. coagulans* is an important spoilage organism affecting milk, vegetables, fruits and pasteurised acid foods, and the thermophilic species *Geobacillus stearothermophilus* (formerly *Bacillus stearothermophilus*) is important in canned foods.

Sources

Environment

Bacilli are natural inhabitants of the soil and water, and are also found in plant matter. Their resistant endospores can be passively distributed very widely and they can be isolated from many sources.

Foods

Bacillus spp can be found mainly in low-acid foods (pH >4.6) that have received a low temperature pasteurisation and are then chilled, or are commercially sterilised (e.g. ultra-high temperature or UHT-treated milk), or acid foods that have received pasteurisation and are stored at ambient temperature. Raw materials such as vegetables, cereals, eggs and meats are also contaminated. The resistance of the endospores means that bacilli are important contaminants of heat-processed and dried foods.

Spoilage Characteristics

The bacilli are involved in a variety of specific types of spoilage in a diverse range of foods. For example, psychrotrophic species produce serious off-flavours such as bitter, putrid, stale, rancid, fruity, yeasty and sour tastes in a wide range of dairy products such as milk, cheese and cream. *Bacillus* spp. such as *B. sphaericus*, *B. subtilis* and *B. lentus* have been associated with flavour defects. They cause problems in pasteurised milks, particularly *B. cereus* strains that produce lecithinase. This enzyme causes a defect known as 'sweet curdling', or 'bitty cream'. 'Flat sour' spoilage of evaporated milk is caused primarily by *Geobacillus stearothermophilus* and *B. coagulans* but can be caused by *B. subtilis*, *B. licheniformis* or *B. macerans*. *B. coagulans* has also been associated with spoilage of evaporated milk, resulting in firm coagulation. It has also been discovered that highly heat-resistant spores belonging to a recently-identified species, *B. sporothermodurans*, may cause spoilage in UHT-processed milks. Psychrotrophic bacilli also cause souring in cooked meat products.

Aciduric, thermoduric species, such as *B. coagulans*, can cause spoilage in acid, heat-processed products such as tomato pastes and sauces, asparagus, corn, lima beans and peas. *B. licheniformis* may grow anaerobically in processed tomato products, causing a localised rise in pH sufficient to allow *Clostridium botulinum* to grow.

Thermophilic species produce 'flat sour' spoilage of low-acid, canned food. *B. subtilis*, *B. pumilus*, *B megaterium* and *B. cereus* are associated with the formation of 'rope', a mucoid, stringy mass, in bread. Rope is characterised by the development of fruity or sweet odours, or rotting fruit/bad cheese. This problem has become more common again recently with the trend away from the use of calcium propionate as a preservative in bread.

B. subtilis is known to produce non-volatile acidity in wines.

Growth/Survival Characteristics of the Organisms in Foods

Temperature

The *Bacillus* species found in foods are able to grow over a wide range of temperatures. Psychrotrophic strains of *B. cereus* are shown to have a lower limit of growth at 2.5 °C, some strains of *B. circulans* are able to grow at 2 °C and *B. macerans*/*B. subtilis* are reported to have a lower limit of growth at 0 °C. Many mesophilic species are able to grow at 50 °C, and some thermophiles grow at 65 °C or higher, but show little growth at 35 °C.

Heat resistance of endospores is variable, with the psychrotrophic strains showing less resistance than mesophilic or thermophilic strains. For example, psychrotrophic *B. cereus* isolates record D-values of 5-8 min at 90 °C, but mesophilic strains show similar values at 100 °C. At 100 °C, a *B. subtilis* strain showed a D-value of 11 min and *B. coagulans* isolates varied from 0.7-7.0 min, dependent on pH. Spores of *Geobacillus stearothermophilus* are very heat-resistant, and D-values of 1.0-5.8 min at 121 °C have been recorded.

pH

Growth of aciduric species, such as *B. coagulans,* has been recorded at pH 3.8, but minimum values of >4.0 are more common for these organisms. *B. cereus* has been reported to grow at pH values between 4.3 and 9.3 and *B. subtilis* from 4.5-8.5.

Water activity

Most *Bacillus* spp do not grow below a_w values of 0.93-0.95, although some strains of *B. subtilis* have been shown to grow at 0.90. Endospores are extremely resistant to drying and survive for very long periods. Salt tolerance is variable, and may be between 2.5% and 10% for species associated with foods.

Atmosphere

Facultatively anaerobic species such as *B. licheniformis* grow well in anaerobic environments, but even aerobic species may show limited growth in sealed containers. Many species are strongly inhibited by carbon dioxide.

Preservatives and biocides

The endospores are generally quite resistant to sanitisers and biocides and are difficult to eradicate from processing environments. *B. subtilis* in bread is inhibited by propionate, but acetic acid is less effective. Bacilli show variable sensitivity to other food preservatives.

Control

Bacillus spp are very common and widely distributed in foods, but tend not to compete well with other spoilage microflora in fresh foods. They are more likely to cause problems in heat-processed foods and products relying on hurdles such as reduced pH and water activity for their stability. Control of flat sour spoilage in canned foods is achieved by minimising cooling times and storage temperatures. Monitoring of ingredients quality is an important control for *Bacillus* spoilage, as is adequate temperature control in chilled foods and good hygiene practice. The bacteriocin nisin is active against bacilli, and may be useful in canned vegetable products.

Radiation doses between 10-50 kGy are required to inactivate bacterial spores, with typical D values of 1-4 kGy, depending on the organism. The D-values for vegetative organisms are generally below 0.8 kGy.

BROCHOTHRIX

Brochothrix spp are related to *Listeria* and to *Corynebacteria*. The genus was proposed in 1976 to accommodate organisms previously classified as *Microbacterium thermosphacta*.

The Organism *Brochothrix*

Brochothrix spp are Gram-positive, non-motile, non-sporeforming rods, but become pleomorphic in older cultures, that occur either singly or in chains . They are facultative anaerobes, and are catalase-positive.

Key Species in Foods

There are two species in the genus, but only *B. thermosphacta* is significant in foods. *B. campestris* is normally found in soil and water.

Sources

Environment

The sources of *B. thermosphacta* are not well understood, but it probably comes from soil and water. It possibly also comes from animal faeces as it is a common inhabitant of the intestinal tract of animals. It is a very common contaminant in meat-processing facilities and can be readily isolated from plant and equipment.

Foods

B. thermosphacta is commonly found in meat and poultry products, but has also been isolated from MAP fish, frozen vegetables and dairy products. The organism is a particular problem in chilled, vacuum-packed and MAP fresh meat products.

Spoilage Characteristics

Spoilage of fresh and cured meat may occur in both aerobic and anaerobic conditions, and is usually apparent as an 'off'-odour, sometimes described as a sour, sweaty or acidic, malty, dairy or cheesy smell. It results from the formation of a variety of metabolic by-products from glucose fermentation, including acetoin, diacetyl, and traces of iso-butyric and iso-valeric acids. Spoilage is often not apparent until some time after the population has reached its maximum level of about 10^5 cfu/g. The production of lipase and protease are other factors contributing to food spoilage. *B. thermosphacta* utilises amino acids, but H_2S is not produced. The organism may grow on fat surfaces, especially in pork and lamb. Lipases are not produced in detectable amounts at temperatures below 20 °C. Production of proteases has been found to be less at 6 °C than at 10 °C.The ecology of spoilage is complex and is affected by pH, gaseous atmosphere and the presence of competing microflora.

Growth/Survival Characteristics of the Organisms in Foods

Temperature

B. thermosphacta is a psychrotroph, able to grow at temperatures as low as -0.8 °C. Most isolates will not grow above 30 °C and have an optimum of 20-25 °C. However, growth may be quite rapid at temperatures above 5 °C.

The organism is not heat-resistant and will not survive even mild pasteurisation treatments. D-values at 55 °C of 1 min have been recorded in skimmed milk and buffers, although resistance is likely to be higher in meats.

pH

B. thermosphacta is favoured by high pH values and has an optimum pH for growth of 7.0, within a range of 5.0-9.0. Spoilage of high-pH meat (>6.5) is much more rapid than for more acid substrates.

Water activity

B. thermosphacta has been reported to grow at a_w values down to 0.94 at temperatures of 20-25 °C in an aerobic atmosphere, and some strains are able to tolerate 10% salt.

Atmosphere

The organism is a facultative anaerobe and will grow in vacuum-packed meats, but it grows more rapidly in air. It will also grow in MAP packs containing elevated oxygen or carbon dioxide levels. It can tolerate up to 50% carbon dioxide, provided some oxygen is present.

Preservatives and biocides

B. thermosphacta can tolerate 100 ppm nitrite at >pH 5.5 and 5 °C aerobically in the presence of 2-4% salt. Some resistance to sulphur dioxide has been reported. Immediate reductions in bacterial counts have been observed in meat products tested with nisin (when combined with lysozyme and EDTA in a gelatin coating).

Control

Since it is so widespread in meat plants and therefore likely to contaminate meat products during processing, effective cleaning and sanitation procedures are essential to control *B. thermosphacta* in MAP and vacuum-packed fresh meats. In addition, as it requires a small level of oxygen to be present, its growth could be inhibited in MAP products where oxygen scavengers are used to reduce the level of carbon dioxide to below 0.2%. Its relative sensitivity to heat means that it is rarely a problem in cooked meats, providing that significant recontamination does not occur. It should be inactivated by a heat treatment of 63 °C for 5 min. *B. thermosphacta* is sensitive to radiation of 2.5 kGy.

CLOSTRIDIUM

The genus *Clostridium* contains a large number of species, many of which are important in foods, both as causes of foodborne disease and as spoilage organisms.

The Organism *Clostridium*

They are Gram-positive, motile or non-motile rods, producing oval or spherical endospores that usually distend the cell. They are catalase-negative, obligate anaerobes, although oxygen tolerance varies widely. Most species have a metabolism that is predominantly either proteolytic or saccharolytic, and there are both psychrophilic and thermophilic species.

Key Species in Foods

Those species responsible for foodborne disease, *C. botulinum* and *C. perfringens*, are not considered here although they can cause spoilage of foods e.g. produce gas and have proteolytic activity.

Many species may be involved in food spoilage, but some of the most significant are *C. sporogenes*, *C. butyricum*, and *C. tyrobutyricum*. These are considered non-pathogenic. *C. butyricum* and *C. pasteurianum* are referred to as butyric anaerobes. Other mesophilic clostridia that have been associated with spoilage of foods include *C. barati*, *C. bifermentans*, *C. beijerinckii*, *C. pseudotetanicum*, *C. fallax*, *C. sordellii* and *C. felsinium*.

Note that there are reports of botulinum neurotoxin production and associated illness in *C. butyricum* and *C. barati*; for more information see the *C. botulinum* chapter.

Some recently recognised psychrotrophic and psychrophilic species, often referred to as 'blown pack clostridia', such as *C. laramie*, *C. aligidicarnis* and *C. estertheticum*, seem to be important in certain types of vacuum-packed meat spoilage. The thermophilic species *C. thermosaccharolyticum*, important in low acid canned food spoilage, has recently been reclassified as a species of *Thermoanaerobacterium*, and the organism still sometimes called *C. nigrificans*, which causes 'sulphide stinker' spoilage in canned foods, is more accurately known as *Desulphotomaculum nigrificans*.

Sources

Environment

Clostridia are ubiquitous in the environment and are found in soil, decaying vegetable matter, animal faeces and marine sediments. The spores are persistent and may be passively distributed.

Foods

Clostridia are present as contaminants in many foods, particularly meat and dairy products. Their heat-resistant endospores may survive in heat-processed foods, and may be able to germinate and grow in sealed packs where anaerobic conditions develop.

C. sporogenes is responsible for causing spoilage of a number of foods including cheese, condensed milk, cooked meat, canned vegetables and canned fish. It is a common *Clostridium* species found in milk and meat. *C. butyricum* is responsible for causing spoilage of cheese, condensed milk and acid canned foods, in particular tomatoes and other fruits (e.g. pear and apricot juices, and canned pineapples) together with *C. pasteurianum*.

C. butyricum is also associated with spoilage of wines. *D. nigrificans* causes spoilage of canned sweet corn, peas, mushroom, mushroom-containing high pH foods and canned baby clams. *T. thermosaccharolyticum* can cause spoilage of acid and acidified products such as tomato and other fruit-containing products, including those containing starchy (farinaceous) ingredients, such as spaghetti. Other products affected include canned sweet potatoes, pumpkin, green beans, mushrooms, asparagus, and vegetable soup.

Spoilage Characteristics

The type of spoilage depends very much on the product and the species involved. Spoilage is manifested by changes in product pH (caused by organic acid production), gas production and production of foul odours (e.g. through volatile acids).

C. sporogenes, the only proteolytic and lipolytic clostridia associated with spoilage, may cause putrefactive spoilage in underprocessed canned foods, and the thermophilic species *T. thermosaccharolyticum* can cause hard swells in cans cooled too slowly or stored at high temperatures. Resulting products are found to be sour, fermented, or have a butyric odour. Putrefactive spoilage of large meat joints may occur if clostridia are present in deep tissue, and grow before the temperature is adequately reduced, or before curing salt concentrations have reached inhibitory levels. The condition known as 'bone taint' in meat carcases is also thought to be caused by growth of clostridia in deep tissue. Psychrotrophic and psychrophilic clostridia (*C. laramie*, *C. aligidicarnis* and *C. estertheticum*) have been found to cause spoilage of chilled, vacuum-packed fresh and cooked meats, producing gas and/or offensive odours. *D. nigrificans* causes black spots on cured meats. *C. pasteurianum*, *C. butyricum*, *C. sporogens* and *C. tyrobutyricum* cause late gassiness of cheese. Of these, *C. tyrobutyricum* is the principal species involved in 'late blowing' of brine-salted hard and semi-hard cheeses, as a result of butyric acid and gas production. *D. nigrificans* cause production of hydrogen sulphide, sometimes resulting in blackening of the product.

Growth/Survival Characteristics of the Organisms in Foods

Temperature

The clostridia are able to grow over a wide range of temperatures. Most are mesophilic, with optimum growth temperatures of 30-37 °C, and are able to grow within the range 10-50 °C. Some are psychrotrophic (maximum growth temperature between 32 °C and 40 °C) and able to grow slowly at 4 °C. The psychrophilic species *C. laramie* (maximum growth temperature below 20 °C) may be able to grow at temperatures as low as -3 °C, and has an optimum of 15 °C and a maximum of 21 °C; it has been found to cause spoilage of meat stored at 2 °C. Thermophilic species have optimum growth temperatures of 50-60 °C; they may grow at 65 °C or higher, but little, if any, growth occurs at 37 °C.

The heat resistance of endospores varies widely. For example, *C. butyricum* spores have a reported D value at 80 °C of 5 min ($D_{100\,°C}$ of 0.1-0.5 min), whereas an extensively studied isolate of *C. sporogenes*, PA3679, has a D value at 121 °C of 0.1-1.5 min ($D_{100\,°C}$ of 80-100 min, z value of 9-13 °C). *T. thermosaccharolyticum* spores are extremely heat-resistant, and D values of 3-4 min have been recorded at 121 °C ($D_{100\,°C}$ of 400 min, z value of 12-18 °C).

pH

Growth at pH values as low as 3.55 have been recorded for species such as *C. butyricum* and *C. pasteurianum* (butyric anaerobes), but minima of > 4.2 are more common for these organisms. The majority of species are inhibited by pH values below 4.5; however, *C. butyricum* and *C. pasteurianium* are capable of growing at pH 4.0.

Water activity

Most clostridia are not able to grow in low-moisture foods, and a_w of 0.93-0.94 would prevent growth of most species. Some are able to tolerate salt concentrations of up to 10%.

Atmosphere

Although the clostridia are obligate anaerobes, some species may tolerate low oxygen concentrations and demonstrate growth in environments that would not normally be considered anaerobic.

Preservatives and biocides

The endospores of clostridia are relatively resistant to many sanitisers and biocides. Nitrites have been used to control clostridia in meat products and cheese, and a bacteriocin, nisin, has also been used to control 'late blowing' in cheese. Phosphates and antimicrobial enzymes like lysozyme are also found to be active against clostridia.

Control

In heat-processed foods that are intended to be commercially sterile, the thermal process is often calculated to destroy the spores of spoilage species, such as *C. sporogenes*, that are heat-resistant. The spores of thermophilic species may be too resistant to control simply by a thermal process, but sufficiently rapid cooling and storage at temperatures below 30 °C will prevent their growth. The risk of thermophilic spoilage can also be reduced by applying strict microbiological standards to ingredients likely to contain thermophilic spores, such as sugars and starches. In chilled foods, adequate temperature control is important, as are quality of raw materials and good hygiene practice. In meat processing, it is important to avoid deep incisions or punctures that may inoculate spores into deep tissues.

Dose required for 90% inactivation of *Clostridium* spores is 3.5 kGy. Radiation values of 1.5 kGy have been found to destroy spores of *C. butyricum*.

ENTEROCOCCUS

The genus *Enterococcus* was formed in 1984 and replaced an earlier classification based on the classical 'faecal streptococci' and Lancefield serological group D streptococci. Nineteen species of *Enterococcus* are currently recognised, but not all of them are of faecal origin and many species are rarely found in foods.

The significance of enterococci in foods is the subject of debate. Some strains are opportunist pathogens, and may cause foodborne disease, though very rarely. They are also cited as indicator organisms (*E. faecalis*, *E. faecium*, *E. durans* and *E. hirae*), but their presence in food is usually not a consequence of faecal contamination. They are sometimes used as starter cultures in food production or as a probiotic. However, some species are important spoilage organisms, and this aspect is considered below.

The Organism *Enterococcus*

Enterococci are Gram-positive, non-sporeforming ovoid cells, occurring singly or in pairs and short chains. Within the chains, the cells are frequently arranged in pairs and are elongated in the direction of the chain. Those species found in foods are non-motile (with the exception of *E. gallinarium* and *E. casseliflavus*). They are facultatively anaerobic and catalase-negative, and have a homofermentative carbohydrate metabolism, producing mainly lactic acid. They have complex nutritional requirements and are resistant to environmental stress. They are easily distinguished from other Gram-positive, catalase-negative homofermentative cocci, such as streptococci and lactococci, in that they are able to grow at 10 °C and 45 °C, in 6.5% NaCl, in the presence of 40% bile at pH 9.6.

Key Species in Foods

By far the most common species found in foods are *E. faecalis* and *E. faecium*. *E. durans* is also occasionally found in milk, and *E. casseliflavus* has been reported to cause yellow discoloration in vacuum-packed cooked meats.

Sources

Environment

Enterococci are found in soil, surface waters, sea water, waste waters and municipal water treatment plants, on plants and in the gastrointestinal tract of animals. *E. faecalis* and *E. faecium* are principally associated with the digestive tracts of humans and other mammals, including poultry, pigs and cattle whereas yellow-pigmented *E. mundtii* and *E. casseliflavus* are typically associated with plants. The natural distribution of other species is not well understood.

Foods

Due to their intestinal habitat in food animals, Enterococci are common contaminants of milk and fresh /processed meat, and may be isolated from numerous other foods. They are also present on fresh produce such as celery, cilantro, mustard greens, spinach, collards, parsley, dill, cabbage and cantaloupe, and possibly originate from the use of untreated irrigation water or manure slurry. They are used as starter organisms in some traditional southern European cheeses.

Spoilage Characteristics

Spoilage usually takes the form of souring as a result of lactic acid production during growth. This may occur in pasteurised milk and dairy products, and in meats, particularly cooked, cured meats. Enterococci may also produce surface slime and cause greening in the core of some cooked meats. They may be involved in the production of biogenic amines such as histamine in some foods.

Growth/Survival Characteristics of the Organisms in Foods

Temperature

The enterococci found in foods are able to grow over a wide range of temperatures. Many strains are psychrotrophic and some isolates of *E. faecium* have been shown to grow at temperatures as low as 1 ºC. Maximum growth temperature has been reported as around 50 ºC, with an optimum for most strains of approximately 37 ºC. These organisms are also notably resistant to freezing.

The enterococci are well known for their heat resistance and are able to survive mild pasteurisation processes. *E. faecium* is notably more heat-resistant than *E. faecalis*, and D values at 70 ºC of 1.4-3.4 min, and 0.02-0.6

min, respectively, have been obtained. *E. faecium* has been associated with spoilage of pasteurised canned ham.

pH

Enterococcus spp are able to grow over a wider pH range than most bacteria, and growth between 4.4 and 10.6 has been recorded.

Water activity

The minimum a_w value for growth varies with the type of solute present, but is reported as 0.93 for *E. faecalis*. Most strains are able to tolerate salt at a concentration of 10% and are resistant to drying.

Atmosphere

Enterococci are facultative anaerobes, but those found in food, especially *E. faecalis*, are well adapted to grow in aerobic conditions.

Preservatives and biocides

The enterococci, particularly *E. faecium*, are generally quite resistant to environmental stress, and therefore persistent, but are not especially resistant to preservatives and biocides. However, multiple antibiotic resistance in isolates of these organisms is causing increasing concern.

Control

Enterococci are commonly present in milk and meat products, and their heat resistance means that they can be a particular problem in pasteurised foods. It is therefore important to ensure that such products receive an adequate heat process. Their persistence and ability to grow in a wide range of environments also mean that they may be able to colonise plant and equipment. Thorough cleaning and sanitising regimes are therefore important in their control.

FLAVOBACTERIUM

The genus was first named in 1923 to describe bacteria forming yellow (*Flavus*) or orange pigmented colonies in culture media.

The Organism *Flavobacterium*

The genus *Flavobacterium* is very heterogeneous, and contains a variety of organisms with certain characteristics in common. They are all Gram-negative, aerobic, non-motile, non-sporeforming, non-fermentative rods *Flavobacterium* spp are catalase- and oxidase-positive, and usually have low nitrogen requirements. Recent revisions have created new genera (*Weeksella*, *Bergeyella*, *Chryseobacterium* and *Empedobacter*), into which many former flavobacteria have been placed, but their importance in foods is uncertain. The genus *Flavobacterium* currently contains 56 species.

Key Species in Foods

Flavobacteria are difficult to characterise, and the position of many is taxonomically uncertain. Therefore, isolates from foods are rarely identified to species level. *Flavobacterium aquatile* is the type species. Species reported in foods include *F. breve* and *F. odoratum*.

Sources

Environment

Flavobacterium spp are ubiquitous and may be easily isolated from fresh water and soil. They are often associated with plant material, and may also be part of the normal skin microflora of animals.

Foods

These organisms have been isolated from a wide variety of chilled foods, including fresh meat and poultry, unpasteurised milk, fish and shellfish, uncarbonated mineral waters and alcoholic beverages produced from grain, where they are a component of the autochthonous or indigenous flora. They are also commonly found in injection brines for cured meat products. They are able to grow and cause spoilage in margarines, emulsions and butter due to their lipolytic properties. Other dairy products like rice pudding, cream, cheddar cheese and milk-based canned goods are also susceptible to spoilage by *Flavobacterium*.

Spoilage Characteristics

Flavobacterium spp are rarely the major cause of spoilage but may be components in the spoilage flora in some foods, such as fish and shellfish, eggs, and vegetable products. Some strains produce pectinases causing degradation of plant material, and heat-stable proteinases and lipases may be produced in milk. They may be involved in the spoilage of pasteurised milk and egg as post-process contaminants. Strains identified as *F. maloloris* were reported to produce a surface taint and 'apple or putrid cheesy odour' in butter, possibly as a result of growth in the cream used in its production. Certain strains may also cause bitterness in dairy products as a result of phospholipase production. They give a vinegary flavour and off-odour to alcoholic beverages.

Growth/Survival Characteristics of the Organisms in Foods

Temperature

Many strains are psychrotrophic and growth at 5 °C or below may occur. The maximum temperature for growth is usually about 30 °C, but some strains will grow at 37 °C. Some isolates are reported to remain viable for long periods in frozen foods.

Flavobacterium spp are normally considered heat-sensitive and are not usually expected to survive pasteurisation. However, there have been reports of survival in pasteurised egg, and D-values in skimmed milk at 63 °C of 4.3-9.6 min were recorded for strains isolated from butter.

pH

There is little published information on minimum values permitting growth. They are not reported to be capable of growth at low pH values, and probably have an optimum pH of 6.0-7.0. Only 83% reduction on levels of *Flavobacterium* was observed in biofilms on stainless steel after exposure to a solution of water at pH11 for 20 min.

Water activity

There is very little information on minimum a_w values that will support growth of *Flavobacterium* spp. Reports of salt tolerance are variable; some strains are inhibited by a concentration of 1%, but others are normally found in meat injection brines or associated with sea fish.

Atmosphere

Flavobacterium spp are aerobic organisms and are normally inhibited by anaerobic conditions. Some strains have been reported to be sensitive to elevated concentrations of carbon dioxide.

Preservatives and biocides

There is very little published information on the effect of antimicrobials on *Flavobacterium* spp, but individual strains have been found to be inhibited by nitrite at a concentration of 100-200 ppm at pH 6. Destruction by chlorine is reported to be rapid.

Control

Flavobacterium spp are normally present on the surface of fresh meat and poultry carcases, but usually die out during storage. They are also likely to be present in milk, fish and vegetables, and have occasionally been shown to become established in food-processing environments. Therefore, effective hygiene and sanitation procedures are important in the control of spoilage. These organisms can also be controlled by pasteurisation, but adequate post-process hygiene is important to prevent recontamination.

A 4 log reduction in levels of *Flavobacterium* on stainless steel has been achieved by exposure to 200 ppm solution of chlorine adjusted to pH 6.5 for 20 min. *Flavobacterium* are also found to be sensitive to essential oils compounds.

HAFNIA

The Organism *Hafnia*

Hafnia spp are members of the Enterobacteriaceae. *Hafnia* is a motile, Gram-negative, non-sporeforming rod. They are facultative anaerobes, possessing both respiratory and fermentative metabolism, and both acid and gas are produced from glucose.

Key Species in Foods

There is currently only one species in the genus, *H. alvei.*

Sources

Environment

H. alvei is present in animal and human faeces, but is not always an indicator of faecal contamination. It can be isolated from both soil and water.

Foods

The organism is found in a wide range of refrigerated foods, particularly meat, fish and seafood, and vegetables. It is also found in eggs and milk.

Spoilage Characteristics

Gaseous and putrefactive spoilage of chilled vacuum-packed meats has been reported, and hydrogen sulphide may be produced. *H. alvei* may also be associated with the production of histamine in scombroid fish, and with spoilage of eggs.

Growth/Survival Characteristics of the Organisms in Foods

Temperature

Some strains are psychrotrophic and growth may be quite rapid at 7 °C. However, reports of minimum growth temperatures are variable; for example, a value of 2.5 °C has been obtained in laboratory media, but the minimum for growth in foods is uncertain. Rapid growth at 43 °C has also been reported.

Heat resistance is not particularly significant. A D-value at 55 °C of 0.36 min has been recorded in peptone water, with a value of 0.74 min being found in smoked fish.

pH

There is little published information on the effect of pH on the growth of *H. alvei*. However, it would be reasonable to expect the organism to grow over a range similar to that of other Enterobacteriaceae (4.5-9.0 is not untypical).

Water activity

There is little published information on the effect of water activity on growth, but related organisms tend to grow at minimum a_w value of approximately 0.95.

Atmosphere

The organism is a facultative anaerobe and will grow in vacuum-packed meats. Under aerobic conditions, *H. alvei* does not compete well with obligate aerobes.

Preservatives and biocides

H. alvei has been reported to be relatively resistant to sulphur dioxide used as a preservative in fresh sausage, but there is little published information on its general resistance to antimicrobial compounds.

Control

The importance of *H. alvei* in spoilage is limited mainly to vacuum-packed meat products. Effective cleaning and hygiene procedures are important to minimise the initial level of contamination of these products. Mild heat processes are also effective.

LACTOBACILLUS

Lactobacillus is a highly heterogeneous genus that contains over 50 species, many of which are important in food spoilage. Some of these species have been placed in a new genus, *Weissella*, including *L. viridescens*, a species that causes greening in meat.

The Organism *Lactobacillus*

The lactobacilli are Gram-positive, non-sporeforming rods or coccobacilli, and are usually non-motile and catalase-negative. They are strictly fermentative, aerotolerant or anaerobic, and aciduric or acidophilic, and usually have complex nutritional requirements. Their carbohydrate metabolism may be homofermentative, producing mainly lactic acid, or heterofermentative, producing a mixture of lactic and acetic acids and carbon dioxide. They are able to grow over a wide range of temperature and may tolerate apparently hostile environments. Some species produce bacteriocins that have an antagonistic effect on other bacteria.

Key Species in Foods

Many species have been isolated from foods, as spoilage organisms, but also as components of the natural microflora of fermented foods. Some species are used as starter cultures in fermented dairy, meat and vegetable products.

They are divided into three groups based on fermentative features. These include obligately homofermentative species such as *L. acidophilus* and *L. delbrueckii*, facultatively heterofermentative species such as *L. casei*, *L. plantarum* and *L. sake*, and the obligately heterofermentative species *L. brevis*, *L. buchneri* and *L. fermentum*.

Sources

Environment

Most *Lactobacillus* spp important in foods probably originate from plant material. They are usually associated with environments containing large amounts of fermentable carbohydrate. Some truly anaerobic species are associated with the mucosal membranes and digestive tracts of animals, but these are rarely found in foods.

Foods

Lactobacilli are found in a variety of foods, particularly dairy products, vacuum-packed or MAP meats and poultry products, alcoholic beverages, canned and acidified or fermented fruit and vegetable products.

Spoilage Characteristics

Atypical flavours, such as cheesy, sour, acid and sometimes liver-like have occurred due to acid formation. Excessive acid production in vacuum-packed cooked meat products can affect sensory attributes. Acids produced can also cause sour/acid odours in raw milk and canned vegetables, bitter flavours in cheese, unpleasant buttery flavours in juices and haziness in alcoholic beverages.

Greening of meat may also occur due to the production of hydrogen peroxide by some species of *Lactobacillus*. Some Lactobacilli may cause or accelerate spoilage by formation of hydrogen sulphide gas which results in malodorous off-odours and black spots on meat. Pigment production, such as pink and orange pigments, can occasionally cause spoilage in cheese.

Some lactobacilli are associated with gas production. They have been associated with the rapid development of carbon dioxide in vacuum packed meat products. Canned vegetables may be spoiled by lactobacilli forming considerable amounts of carbon dioxide, causing blown cans. Blown cans can also occur with sour fruit products. In salad dressings the gas formed leads to bubble-filled dressing and/or blowing of the containers. Bloater damage in brined cucumbers results from an increase in gas pressure inside the cucumbers during fermentation. 'Protein swell' is an unusual type of spoilage caused by lactobacilli. It raises the pH (due to ammonia production) of the spoiled product. In this case, the proteins are decomposed and the subsequent decarboxylation of amino acids leads to enhanced gas production.

Exopolysaccharides (EPS) can be produced in meat products, cider and wine causing severe problems. *Lactobacillus sake* has been found to cause 'ropy' spoilage' of vacuum-packed cooked meat products, by producing the EPS dextran and H_2S. Heterofermentative lactobacilli have been associated with the ropy spoilage of wine and/or cider.

Growth/Survival Characteristics of the Organisms in Foods

Temperature

Lactobacillus spp are able to grow over a wide temperature range. Some species are psychrotrophic and may grow at temperatures as low as 2 °C and some are moderately thermotolerant, growing at temperatures up to 55 °C. Most species have optimum growth temperatures of 30-40 °C. Thus growth may occur in chilled and ambient foods.

Lactobacilli are relatively heat-resistant and may survive inadequate pasteurisation processes. For example, D-values as high as 11 min at 70 °C have been reported for *L. plantarum*. Heat resistance varies with pH, a_w and other factors, and may be significantly higher in low-moisture foods.

pH

Lactobacilli are all acid-tolerant, but minimum pH for growth varies with species and with the acid present. *L. brevis* will grow at pH 3.16 in the laboratory, but *L. lactis* does not grow below 4.3. *L. suebicus* isolated from fermented apples and pear mashes and *L. acetotolerans* from rice vinegar are able to grow at pH 2.8 and pH 3.3 respectively. Most species are inhibited by pH values above neutral.

Water activity

The minimum a_w value for growth varies with species and with the type of solute present, but is likely to be within the range 0.93-0.96. Lactobacilli are not particularly sensitive to dry conditions and survive for long periods in dried vegetables.

Atmosphere

Most of the species important in foods are aerotolerant, but do not grow well in aerobic environments. Sufficient growth to cause spoilage is unlikely to occur unless the redox potential is reduced.

Preservatives and biocides

Some spoilage species are quite resistant to certain food preservatives, particularly acetic acid, nitrites/nitrates, and sulphur dioxide. *L. suebicus* is able to tolerate ethanol concentrations of 12-16%.

Control

Lactobacilli are very common contaminants in plant material and are often present in meat and dairy products. They are one of the few spoilage organisms that cannot easily be controlled by reducing pH, and may grow well in environments that cannot be readily colonised by other bacteria. Spoilage usually becomes detectable only slowly, especially at low temperatures, and therefore adequate temperature control of susceptible products is important in achieving the desired shelf-life. Adequate heat processing, satisfactory microbiological quality of ingredients and good hygiene practice are all important controls.

Note: A recent study has shown that *Lactobacillus hilgardii* are spoilage, high-histamine-producing bacteria in wines. Histamine is a biogenic amine that can cause headaches and allergic-like reactions in susceptible individuals.

LEUCONOSTOC

The Organism *Leuconostoc*

Leuconostoc spp are lactic acid bacteria and can often be found in association with *Lactobacillus* spp. They are Gram-positive, non-sporeforming, non-motile, catalase-negative, heterofermentative cocci. They are dependent for growth on the presence of fermentable carbohydrate, and have complex nutritional requirements. End products from glucose metabolism are lactate, ethanol and carbon dioxide. Some strains produce an extracellular polysaccharide, dextran, from sucrose. *Leuconostoc* spp are used as starter cultures in certain fermented dairy and vegetable products.

Key Species in Foods

The principal species in foods is *L. mesenteroides*. Three subspecies of this organism have been recognised - *mesenteroides*, *dextranicum* and *cremoris* (usually associated with dairy products). Other species of importance in food spoilage are *L. gelidum* and *L. carnosum*, which are associated with meat products, *L. gasicomitatum* associated with spoilage of acetic acid fish preserve, *L. paramesenteroides* (recently placed in the genus *Weissela*) and *L. oenos*, a species associated with wine and now reclassified as *Oenococcus oeni*.

Sources

Environment

The primary environmental source of *Leuconostoc* spp is plant material. They are reported to be the commonest type of lactic acid bacteria on growing vegetation. Other sources to foods include food utensils, gastrointestinal tract and animal hides.

Foods

Leuconostoc spp are often associated with meat products, particularly vacuum-packed or MAP sliced meats, dairy products, and fermented vegetable foods, such as sauerkraut. *L. mesenteroides* strains are common contaminants of sugar-processing operations. It has also been responsible for spoilage of wine and cider. *L. carnosum* is reported to be strongly associated with vacuum-packed ham.

Spoilage Characteristics

Leuconostocs are rarely the dominant organisms in spoiled products, but they may contribute to spoilage in a variety of foods. *L. mesenteroides* subsp. *mesenteroides* and *dextranicus* are involved in souring and blowing of vacuum-packed or MAP meats, and, where exopolysaccharides are produced, a 'ropy slime' spoilage may result. *L. carnosum* is associated with spoilage of vacuum-packed ham. Acid and dextran production by *L. mesenteroides* subsp. *mesenteroides* can cause serious problems in sugar processing, and sufficient dextran may be produced to block pipelines. Some strains may cause blowing in certain cheeses, and spoilage of concentrated orange juice has also been reported.

Growth/Survival Characteristics of the Organisms in Foods

Temperature

Some *Leuconostoc* strains are psychrotrophic and are able to grow at 2 °C. *L. carnosum* is able to grow slowly at 1 °C, but not at 37 °C. A minimum growth temperature of 5-8 °C is more common; with maxima in the range 40-50 °C. Optimum temperature for dextran production is usually 20-25 °C.

The heat resistance of *Leuconostoc* spp varies, but most strains are not notably heat-resistant. A D-value at 60 °C of 31.3 s has been reported for *L. mesenteroides*, but strains that produce extracellular polysaccharides may be notably more resistant and may survive brief exposure to temperatures as high as 85 °C.

pH

Minimum pH for growth varies with species and with acidulant, but a strain of *L. mesenteroides* has been reported to be capable of growth at 3.5. Optimum pH for most species is nearer 6.0, with the exception of *L. oenos*, which is acidophilic and has an optimum of 4.2-4.8.

Water activity

There is little information on the effect of a_w on *Leuconostoc*. Most species are not particularly salt-tolerant, but *L. fallax*, a species found in sauerkraut, has been reported to grow in the presence of 6.5% salt. *L. mesenteroides* subsp. *mesenteroides* grows well in sugar cane juice having a sucrose content of 15-20%.

Atmosphere

Leuconostoc spp are facultative anaerobes and will grow in vacuum-packed and MAP meats. Growth in the presence of oxygen is said to be enhanced by the presence of 15% carbon dioxide.

Preservatives and biocides

L. carnosum, associated with cured meats, has been reported to show some resistance to nitrite, and *L. oenos* is quite resistant to sulphur dioxide. Extracellular dextran production may have a protective effect on cells exposed to biocides.

Control

Leuconostoc spp can be difficult to eradicate from food-processing environments, particularly where large quantities of sugar are present. Those strains that produce dextran from sucrose show enhanced resistance to heat and other environmental stresses, and readily form biofilms in association with other spoilage organisms. For cooked meat products, an adequate heat process is required, and effective cleaning and hygiene procedures are essential to prevent recontamination during operations such as slicing. Experiments using Pulsed Electric fields of 30-50 kV/cm at 50 °C have resulted in 5-log cycle reductions of *Leuconostoc mesenteroides* in orange juice.

MICROCOCCUS

The Organism *Micrococcus*

Micrococcus spp are Gram-positive, non-sporeforming, non-motile cocci that occur in pairs, tetrads or clusters, but not in chains. They are both catalase- and oxidase-positive. Micrococci are generally obligate aerobes with a strictly respiratory metabolism, although one species has been reported to be capable of anaerobic growth. Some species produce pigmented colonies ranging from yellow to pinkish-red. Recently, some species have been transferred to new genera, but only the genus *Kocuria* is associated with foods. The distinction between micrococci and staphylococci is not always clear, and misidentification may occur.

Key Species in Foods

Species isolated from foods include *M. luteus* and *M. varians*. The latter has recently been reclassified as *Kocuria varians*, but is still generally referred to as *M. varians* in the literature. Micrococci isolated from foods are often not identified to species level.

Sources

Environment

The skin of animals appears to be the natural environment of most *Micrococcus* spp; they are also common airborne contaminants and are frequently isolated in dust and on dry surfaces.

Foods

Micrococci are often found associated with dried meat products and are also common in milk and on the surface of cattle and sheep carcases. They may be used as starter cultures in some fermented meat and dairy products.

Spoilage Characteristics

Micrococci are not generally considered to be important spoilage organisms but may contribute to spoilage of a wide range of products, particularly those subjected to pasteurisation, or intermediate-moisture foods. Extensive growth on meat surfaces may cause discoloration, and slime may be formed on the surface of cured meats. They have been isolated from spoiled pasteurised milk and liquid egg.

Growth/Survival Characteristics of the Organisms in Foods

Temperature

Data on the effect of temperature on the growth of micrococci are sparse, but some strains are psychrotophic. The optimum growth temperature is thought to be between 25 °C and 37 °C. *Micrococcus* spp have been reported to grow on meat at 4 °C, and isolates of *M. luteus* and *M. varians* were found to grow in milk at 7.2 °C. A strain of *M. varians* did not grow on MAP meat at 4 °C, but did grow at 10 °C. The same organism grew at 37 °C, but not at 45 °C.

Micrococci can be described as thermoduric and may survive mild heat processes. *M. varians* was found to have a D-value at 70 °C of 0.15 min in a fish terrine, and has been reported to survive in milk at 68 °C for 16 s. An isolate identified as *M. freudenreichii* was found to have a D-value at 62 °C of 3.5 min. Survival at such temperatures can be extended greatly at reduced a_w values and in fatty products, but the effect is less marked at higher temperatures.

pH

There is little information on the effect of pH on growth, but *M. varians* was found to grow at a pH of 5.0, and some isolates of *Micrococcus* spp have been reported to grow at 4.5.

Water activity

Micrococci are quite resistant to drying and may survive for long periods in low-moisture environments. Growth of *M. luteus* has been shown at a_w of 0.93, and some isolates are able to grow at a salt concentration of 15%. Micrococci have also been isolated from curing brines with salt concentrations of approximately 20%.

Atmosphere

Most species are strict aerobes, but *M. varians* has been shown to be capable of growth in the absence of oxygen and may be involved in spoilage of vacuum-packed meats or meat stored under carbon dioxide.

Preservatives and biocides

There is very little published information on the effect of antimicrobials on micrococci; however, a strain of *M. luteus* has been shown to be more resistant to four common sanitisers than several other Gram-positive organisms.

Control

Micrococci are often overlooked in foods, but may be troublesome in pasteurised products, especially those with a high salt content, where there are few competing organisms. They may persist for long periods on dry surfaces, and effective hygiene programmes are necessary to prevent colonisation of suitable environments. Prevention of airborne contamination by micrococci requires adequate air filtration and handling systems.

MORAXELLA

The Organism *Moraxella*

Moraxella spp belong to the family Moraxellaceae along with *Acinetobacter*, *Psychrobacter* and other related groups. They are Gram-negative, aerobic (although some strains are weakly facultative), non-motile, non-sporeforming organisms. The genus *Moraxella* contains two distinct groups or subgenera. In the first group the organisms occur in pairs or chains and are rod-shaped. They are in the subspecies *Moraxella*. The second subgroup, *Branhamella*, contains the cocco-bacilli that occur singly or in pairs. However, they show pleiomorphic characteristics when grown in conditions of reduced oxygen and elevated temperatures. They have a strictly respiratory metabolism, are oxidase- and catalase-positive, and do not produce acid from glucose. They can be distinguished from *Acinetobacter* on the basis of their oxidation reaction.

Key Species in Foods

Species isolated from foods include *M. osloensis*, *M. atlantae*, and *M. phenylpyruvica*, although it has recently been proposed that the last species be transferred to the genus *Psychrobacter*.

Sources

Environment

Moraxella spp are widely distributed in soil. They are also found in marine environments and are present on the skin of humans and animals.

Foods

These organisms are often found on chilled proteinaceous foods such as fresh meat and poultry, and are usually part of the initial microflora on the surface of dressed animal carcases. They have also been isolated from dairy products, and form the predominant flora of fish.

Spoilage Characteristics

Moraxella spp are part of the normal spoilage flora of fresh meat and poultry stored under aerobic conditions. They may produce 'off'-odours as a result of amino acid metabolism, but are often regarded as having low spoilage potential.

Growth/Survival Characteristics of the Organisms in Foods

Temperature

The optimum growth temperature is 33-35 °C. Some strains are psychrotrophic, and growth at 4 °C may occur, but *Moraxella* spp do not compete well under these conditions.

pH

Moraxella spp are favoured by pH values of 6.0 or above, and do not grow well on meats at lower pH.

Water activity

A minimum a_w value permitting growth as high as 0.99 has been reported for a meat-spoilage isolate. *Moraxella* spp tend to die out on the surface of lamb carcases during storage, and it has been suggested that this is due to desiccation.

Atmosphere

Moraxella spp are aerobic organisms and are not normally involved in the spoilage of vacuum-packed or similar products. However, a few strains have been reported to grow slowly under anaerobic conditions.

Preservatives and biocides

There is very little published information on the effect of antimicrobials on *Moraxella* spp.

Control

Moraxella spp are normally present on the surface of fresh meat and poultry carcases, and may be involved in spoilage under aerobic conditions, particularly if the temperature is allowed to rise above 8 °C. Spoilage is therefore best prevented by careful temperature control. Prevention of cross-contamination from the skin during processing by good hygiene practice is also important. Kill values reported for radiation treatment of *Moraxella* spp. is between 0.191 and 0.86 kGy.

PHOTOBACTERIUM

The Organism *Photobacterium*

The genus *Photobacterium* spp belonging to the family Vibrionaceae are large, Gram-negative, non-sporeforming rods that are usually motile; they are capable of fermentative and oxidative metabolism. Oxidase reaction is variable, and most isolates have a positive requirement for sodium ions. Some species show marked bioluminescence.

Key Species in Foods

There are currently 12 species and 2 subspecies of *Photobacterium*; however, *Photobacterium phosphoreum* is the species most often found in foods.

Sources

Environment

The principal source of *Ph. phosphoreum* is temperate marine waters.

Foods

Ph. phosphoreum is generally associated with marine fish, and can be isolated from the skin and in the digestive tract.

Spoilage Characteristics

It has recently been suggested that *Ph. phosphoreum* is the principal organism responsible for spoilage in chilled MAP and vacuum-packed temperate marine fish, particularly cod. Spoilage is usually apparent as objectionable odours, produced by the reduction of trimethyl amine oxide (TMAO) to trimethyl amine (TMA). TMA is largely responsible for the characteristic fishy odour of spoiled fish. The importance of the organism in fish spoilage generally is still uncertain, partly because conventional culture techniques may greatly underestimate its numbers.

In addition to being a spoilage organism, *Photobacterium* has been associated with food poisoning due to histamine production. *Ph. Phosphoreum* was found to be the causative organism in an incidence of food poisoning due to histamine fish poisoning from dried sardines.

Ph. phosphoreum has also been isolated in large numbers from visibly luminescent cooked shellfish. Post-cooking contamination and subsequent growth during storage were thought to have occurred.

Growth/Survival Characteristics of the Organisms in Foods

Temperature

Optimum temperature for growth is generally 18-25 °C. *Ph. phosphoreum* is psychrotrophic and grows well at 5 °C. Minimum temperature for growth is reported to be approximately 0 °C in fish, and growth does not usually occur at 35 °C.

Information on heat resistance is very scarce, but the organism seems to be quite heat-sensitive and may be killed by the temperatures attained in pour plate culture techniques.

pH

There is little published information on the effect of pH, but the minimum value for growth is probably approximately 5.0.

Water activity

There is very little published information about the effect of a_w on the growth of *Ph. phosphoreum*, but growth is reported to occur at a salt concentration of 6.5%.

Atmosphere

The organism is a facultative anaerobe, and growth has been demonstrated in MAP and vacuum-packed fish. It is reported to be carbon dioxide tolerant. Reports have shown that histamine production is strongly inhibited in products packed in 40% carbon dioxide/60% oxygen.

Preservatives and biocides

There is little published information on the effect of biocides and preservatives. However, inhibition of growth in MAP fish by potassium sorbate and chelating agents has been reported.

Control

Ph. phosphoreum appears to be quite common in temperate marine fish and may be more important in spoilage than is appreciated. Its heat sensitivity may allow control by minimal thermal processes. It has also been suggested that specific inhibition of the organism in MAP fish may extend shelf-life significantly.

PROTEUS

The Organism *Proteus*

Proteus spp are members of the Enterobacteriaceae and are Gram-negative, non-sporeforming rods that often display pleomorphism. They are motile and often show swarming growth on moist agar plates. They are facultative anaerobes, possessing both respiratory and fermentative metabolism, and both acid and gas are produced from glucose. Some organisms previously classified as *Proteus* spp have recently been assigned to other genera.

Key Species in Foods

There are now only two species important in foods, *P. mirabilis* and *P. vulgaris*. *P. morganii*, which is an important producer of histamine in scombroid fish, has been placed in the new genus *Morganella*.

Sources

Environment

Proteus spp are found in human and animal faeces, soil and polluted waters.

Foods

These organisms are found in vegetables, meat, poultry, fresh-water fish and raw milk. Their entry into the food chain may be associated with foods via faecal contamination; however, they are not necessarily a reliable indicator of this type of contamination.

Spoilage Characteristics

Proteus spp may be responsible for spoilage of cured meats such as dry-cured hams, sweet-cured bacon, and vacuum-packed bacon. Spoilage may take the form of souring or tainting. The spoilage of vacuum packed bacon is referred to as 'cabbage odour'. In cottage cheese, it causes spoilage termed as 'slimy curd' due to slime formation and production of off-flavours and off-odours. They may also cause 'custard rots' in eggs, and have been shown to produce histamine in scombroid fish.

Growth/Survival Characteristics of the Organisms in Foods

Temperature

These organisms are usually considered to be mesophilic, but a few strains have been shown to grow psychrotrophically. *P. vulgaris* has been found to grow very slowly at 4 °C, but growth has not been detected at 1 °C.

Few data have been published on the heat resistance of *Proteus* spp, but it has not been reported to be very significant.

pH

Growth has been shown to occur over the pH range 4.4-9.2 in laboratory media.

Water activity

There is little published information on the effect of water activity on growth. Growth in cured ham was reported at a salt concentration of 5%, but not at 6%.

Atmosphere

The organism is a facultative anaerobe and will grow in vacuum-packed bacon.

Preservatives and biocides

There is little published information on the effect of antimicrobials.

Control

Proteus spp are only really important as spoilage organisms in chilled foods under conditions of temperature abuse. Therefore, good temperature control is essential. Adequate hygiene procedures are also important controls, especially for cured meat products that are not kept below 5 °C throughout processing.

PSEUDOMONAS

The Organism *Pseudomonas*

The genus *Pseudomonas* is large and heterogeneous, but contains several species that are very important in spoilage, particularly of chilled foods, since they are psychrotrophic. The pseudomonads are Gram-negative, oxidase- and catalase-positive small rods. They are aerobic and the cells are usually motile. Some species produce water-soluble, fluorescent pigments.

Key Species in Foods

Ps. fluorescens and *Ps. fragi* are the two species most commonly associated with spoilage of chilled foods, and both *Ps. lundensis* and *Ps. putida* are also important. *Ps. aeruginosa* is an opportunistic pathogen, but is probably of more significance in water supplies than in foods.

Sources

Environment

Pseudomonas spp are important components of the microflora of soil and water, and are widely distributed in the environment. They are frequently isolated from plant matter.

Foods

Pseudomonads are common contaminants of fresh produce including spinach, lettuce, cabbage, potatoes, and tomatoes, and are usually the predominant spoilage organisms in chilled foods with a near neutral pH and a high water activity, such as fresh meat, poultry, milk and fish.

Other foods with which pseudomonads are associated include butter, cocoa beans, cream, eggs, cereals and mushrooms.

Spoilage Characteristics

The aerobic growth of pseudomonads in chilled foods may result in spoilage due to 'off'-odours produced as the organisms utilise amino acids, releasing sulphur-containing compounds, such as dimethyl sulphide, and amines, such as cadaverine and putrescine. Pseudomonad populations of 10^6 cfu/g may be sufficient to produce odours in milk. In fresh meats, surface counts of $>10^7$ cfu/cm^2 are required for odour production, and populations of 10^8 cfu/cm^2 or more produce visible slime on the meat surface. *Pseudomonas* spp also produce extracellular proteolytic enzymes and lipases, which may have a direct role in spoilage, and which are able to survive some pasteurisation processes and affect products such as UHT milks.

They are also associated with spoilage of butter, resulting in 'off'-odour formation and rancidity. Some are associated with the formation of fruity odours or black discoloration of the butter.

Growth/Survival Characteristics of the Organisms in Foods

Temperature

The species important in food spoilage are usually psychrotrophic, and are able to grow and form colonies at 0-7 °C. For example, *Ps. fragi* has been observed to produce detectable slime on fresh meat stored at 1 °C, and slow growth has been demonstrated for *Ps. fluorescens* at 0 °C. Mesophilic pseudomonads do not grow at 10 °C or below. Maximum growth temperatures vary, but key food-spoilage species do not usually grow above 35-40 °C.

Pseudomonas spp are not heat-resistant and are readily destroyed by mild pasteurisation treatments. D-values at 55 °C in the range of 1-6 min are reported for food-spoilage species.

pH

Pseudomonads are not acid-tolerant and most species are able to grow within a range of 5.0-9.0, with an optimum of 6.0-8.0.

Water activity

Growth does not occur at low a_w values, and pseudomonads are not tolerant of drying. They are found on the surface of foods with a_w 0.99 or higher. No sign of growth has been observed at a_w 0.91 The minimum a_w for growth is dependent on other factors, such as the solute present, pH, etc., but, under otherwise ideal conditions, growth at 0.95-0.97 is possible. However, for practical purposes in foods, values below 0.97 severely inhibit these organisms.

Atmosphere

Pseudomonads are obligate aerobes and do not grow or cause spoilage in the absence of oxygen. Their presence in large numbers in foods that would not be expected to support their growth, such as vacuum-packed meats, may indicate faulty packaging.

Preservatives and biocides

Pseudomonas spp tend to be relatively resistant to many biocides, including quaternary ammonium compounds (QAC). They are particularly difficult to eradicate from contaminated equipment if present within biofilms.

At pH 6 and incubation at 5 °C in nutrient broth, >10 ppm of diacetyl inhibited *P. flourescens*.

Control

Pseudomonads are common, and widely dispersed in foods and the environment. Their ability to grow at low temperatures and cause spoilage of a wide range of chilled foods can cause serious problems for manufacturers. Control of spoilage in susceptible products is best achieved by the use of high-quality raw materials, and the operation of thorough cleaning and hygiene regimes. Mild heat processes are a very effective means of destroying pseudomonads, but high standards of post-process hygiene are necessary to prevent recontamination.

Chemical interventions used to control *Pseudomonas* spp include ozone, chlorine, hydrogen peroxide and organic acids. Studies have shown that exposure of *P. fluorescens* to 2.5 ppm of ozone for 40 s results in a 5-6 log reduction in bacterial counts. 4 log reductions of surface contamination have also been achieved with hydrogen peroxide. *Pseudomonas* spp. is sensitive to chlorine and chlorine dioxide. The D_{10} value for *P. fluorescens* subjected to irradiation is in the range of 0.5-1.0 kGy.

PSYCHROBACTER

The Organism *Psychrobacter*

The genus *Psychrobacter,* belonging to the family Moraxellaceae, was proposed to include organisms previously classified as *Acinetobacter* and *Moraxella*, and also contains those organisms sometimes referred to as *Moraxella*-like. They are Gram-negative, non-motile rods or cocco-bacilli, often occurring in pairs. They are aerobic, catalase- and oxidase-positive, and their metabolism is respiratory.

Key Species in Foods

The genus originally contained only one species, *Psy. immobilis*, first described in 1986; since then 26 species have been described. The importance of these species in foods is as yet unknown.

Sources

Environment

The natural habitat of *Psy. immobilis* is water. It has also been isolated from soil, air and sea ice.

Foods

The organism is commonly isolated from proteinaceous foods such as meats, poultry, fish and milk.

Spoilage Characteristics

Psy. immobilis is part of the typical spoilage microflora found on chilled fresh meats and poultry. Its importance is uncertain, since it is not proteolytic and does not compete well with pseudomonads. However, it does produce some volatile metabolites and may cause lipolysis in meat if present in high numbers. It has also been found as a post-pasteurisation contaminant in spoiled pasteurised milk.

Growth/Survival Characteristics of the Organisms in Foods

Temperature

Psy. immobilis is psychrotrophic and grows well at 5 °C. Minimum temperature for growth is reported to be approximately 1-2 °C, optimum growth temperature is 20 °C. Growth does not usually occur at 35-37 °C and those strains that grow do not normally grow at 5 °C.

Information on the effect of high temperatures is very scarce, but *Psy. immobilis* has not been reported to have any significant heat resistance.

pH

There is little published information on the effect of pH, but growth seems to be favoured by near neutral pH values above 5.5.

Water activity

There is very little published information about the effect of a_w on the growth of *Psy. immobilis*, but growth is reported to occur at a salt concentration of 6.5%.

Atmosphere

The organism is a strict aerobe; however, some strains can grow under anaerobic conditions provided there is a suitable electron acceptor.

Preservatives and biocides

There is little published information on the effect of biocides and preservatives.

Control

Selection of good-quality raw materials and mild heat processes are effective controls for *Psy. immobilis*. However, as with pseudomonads, adequate sanitation and hygiene procedures are vital to prevent recontamination, particularly in dairies.

SERRATIA

The Organism *Serratia*

Serratia spp are members of the Enterobacteriaceae and are motile, Gram-negative, non-sporeforming rods. They are facultative anaerobes, possessing both respiratory and fermentative metabolism, and are catalase-positive. Acid and sometimes gas is produced from glucose. Some strains produce pink or red pigments, but the colonies of others are colourless.

Key Species in Foods

The principal species of importance in foods is *S. liquefaciens*. Other species that may occasionally be isolated are *S. marcescens* and *S. plymuthica*.

Sources

Environment

Serratia spp are found in soil, water and the digestive tracts of animals, and on plant material.

Foods

The organisms are found in a wide range of foods, particularly refrigerated meat and vegetables, and eggs. *S. plymuthica* is often associated with freshwater fish.

Spoilage Characteristics

S. liquefaciens is an important spoilage organism in packed meat products. It produces volatile amines such as putrescine and cadaverine in vacuum-packed meats. It grows very well in vacuum-packed, high-pH meat, and H_2S is produced under these conditions. This is one reason why dark, firm, and dry (DFD) meat is not considered suitable for vacuum packing. *Serratia* spp may also produce heat-stable lipolytic enzymes in dairy products and may be involved in egg spoilage. *S. marcescens* is sometimes associated with a condition known as 'red bread', where growth and pigment production occur within bread that has high moisture content.

Growth/Survival Characteristics of the Organisms in Foods

Temperature

Some strains of *S. liquefaciens* are psychrotrophic and growth occurs at 4-5 °C. However, reports of minimum growth temperatures are variable. A value of 1.7 °C has been obtained in laboratory media, but the minimum for growth in foods is uncertain.

There is little published information on the heat resistance of these organisms, but there are no established reports of strains that are able to survive pasteurisation processes.

pH

Growth of *S. liquefaciens* is favoured by pH values of >6.0 in meats. The pH range within which growth occurs is approximately 5-9.

Water activity

A minimum water activity permitting growth of 0.93 has been reported for *S. liquefaciens*.

Atmosphere

The organism is a facultative anaerobe and will grow in vacuum-packed meats. It will also grow in MAP meats and is resistant to carbon dioxide, having been shown to cause meat spoilage at concentrations of 50%.

Preservatives and biocides

S. liquefaciens is reportedly not inhibited by sorbic or benzoic acids at 0.15% in vacuum-packed meat. However, acetic acid at a concentration of 0.1% did delay growth and higher concentrations prevented it. Studies have shown that *S. liquefaciens* and *S. marcescens* are resistant to 185-270 ppm sulphur dioxide.

Control

The importance of *S. liquefaciens* in spoilage is limited mainly to vacuum-packed and DFD meat products. Effective cleaning and hygiene procedures are important to minimise the initial level of contamination of these products. Mild heat processes are also effective.

SHEWANELLA

The genus *Shewanella* was proposed in 1985, for a group of organisms previously classified as *Alteromonas putrefaciens* and, before that, *Pseudomonas putrefaciens*. Originally, one species, *S. putrefaciens*, was included, but this included a highly heterogeneous group of isolates from different sources, and there are now a further three accepted species.

The Organism *Shewanella*

The organisms are Gram-negative, straight or curved motile rods. *S. putrefaciens* is oxidase-positive and has a respiratory metabolism, but is well able to grow in the absence of oxygen using alternative terminal electron acceptors.

Key Species in Foods

Although the genus *Shewanella* comprises more than 20 species only *S. putrefaciens* is important in food spoilage. The other species are all found in marine environments. There have been suggestions that one of these,

S. alga, may be a human pathogen, but it is not known whether it can cause foodborne infection.

Sources

Environment

The principal source of *S. putrefaciens* is temperate marine waters, but it has been isolated from fresh water. It is also found in oil field wastes, surfaces of processing equipment and clinical specimens.

Foods

S. putrefaciens predominate in marine fish, particularly ice-stored, or MAP fish. However, it can also be isolated from poultry and from other meats, particularly high-pH meat.

Spoilage Characteristics

Spoilage is usually apparent as objectionable odours. In fish, especially gadoid fish such as cod, *S. putrefaciens* reduces trimethyl amine oxide (TMAO) to trimethyl amine (TMA). TMA is largely responsible for the characteristic odour of spoiled fish. The organism also produces sulphur compounds including H_2S, methyl mercaptan, and dimethyl disulfide from sulphur-containing amino acids. *S. putrefaciens* is also a major cause of 'greening' in high-pH meat, and may grow rapidly on chicken leg meat and skin, both of which have higher pH values than breast meat.

Growth/Survival Characteristics of the Organisms in Foods

Temperature

S. putrefaciens is psychrotrophic and capable of forming colonies at 0-7 °C. They grow well at 5 °C. Minimum temperature for growth is reported to be approximately 2 °C. *S. putrefaciens* is unable to grow at temperatures exceeding 37 °C.

Information on heat resistance is very scarce, but an isolate from marine fish was reported not to survive at 55 °C for 20 s.

pH

Growth of *S. putrefaciens* is favoured by pH values above 6.0. It will not tolerate acid conditions, and a minimum pH value for growth of approximately 5.3 has been reported. In meat stored at low temperatures, growth does not usually occur below pH 6.0.

Water activity

S. putrefaciens are found in foods having a_w 0.99 or higher. They show no signs of growth at a_w 0.91. The minimum a_w required for their growth is in the range of 0.95-0.97 depending on the type of food and salts or sugars used. Inhibition of growth by salt at a concentration of 5% has been reported.

Atmosphere

Although principally an aerobic organism, the ability of *S. putrefaciens* to grow anaerobically enables it to grow very well in vacuum-packed and MAP fish and meats. Inhibition by high concentrations of carbon dioxide (up to 10%) has been reported.

Preservatives and biocides

There is little published information on the effect of biocides and preservatives. However, an isolate of *S. putrefaciens* from poultry was found to be more resistant than *Pseudomonas* spp to sodium hypochlorite and other sanitisers.

Control

S. putrefaciens can often be readily isolated from fish- and meat-processing plants, and may become established in such environments. Therefore, effective cleaning and sanitation are important controls to minimise product contamination. Reduced pH is also an effective control in foods. Ionising radiation of 1.5 kGy has been found to be effective in reducing the number of *S. putrefaciens* on the surface of beef, pork, turkey and poultry.

OTHER BACTERIA THAT MAY CAUSE MICROBIOLOGICAL SPOILAGE

The principal genera of bacteria that are involved in the microbiological spoilage of foods are covered in the preceding pages of this section of Micro-Facts. However, this list cannot be exhaustive, and there are other bacteria that have occasionally been implicated in the spoilage of particular foods. Some examples are given below.

Aeromonas

Aeromonas spp are Gram-negative, catalase- and oxidase-positive, facultatively anaerobic rods, that produce large quantities of gas from the fermentation of carbohydrates. Although considered generally aquatic organisms they have been recovered from different animals (beef, chicken, pork, lamb, and raw milk), plants (broccoli, celery, spinach, alfalfa sprouts), and food products of animal origin. They are frequent contaminants of marine water and freshwater, but can also be found in chlorinated and unchlorinated drinking water-distribution systems, and bottled uncarbonated mineral drinking water. Seafood, such as shellfish, oyster and fish, are common sources of this organism. *Aeromonas hydrophila* is an occasional cause of foodborne disease (an enterotoxin has been identified in *A. caviae* and *A. hydrophilia*), but is also important in fish spoilage, particularly at temperatures of >5 °C. Other species are also common, and many are psychrotrophic or psychrophilic. *Aeromonas* spp are important in the spoilage of freshwater and farmed fish, and some processed fish products.

Alcaligenes

The genus *Alcaligenes* consists of Gram-negative, aerobic catalase and oxidase positive, non-fermentative rods, which may produce alkali from certain substrates. Some strains are capable of anaerobic respiration in the presence of nitrite or nitrate. They can grow between 5 °C and 37 °C but have an optimum growth temperature of between 20-37 °C. They are found in soil and water, and are common inhabitants of the intestinal tract of some animals. They are therefore considered potential contaminants of dairy products, meat and poultry. They have been found in bottled and ground spring waters, marine and fresh fish, and on raw vegetables. *Alcaligenes* spp have been associated with spoilage of eggs, pasteurised milk as a post-process contaminant, some meat products, such as bacon, and with rancidity in butter.

Carnobacterium

The genus *Carnobacterium* was proposed in 1987 to include species of lactobacilli. They are Gram-positive, catalase-negative rods. These organisms are psychrotrophic, and heterofermentative, but are unable to grow at pH 4.5, unlike most lactobacilli. This non-aciduric genus is highly specific to fresh meat and mainly poultry. *C. divergens* is associated with vacuum-packed meat products and may produce tyramine (a product formed due to the decarboxylation of the amino acid tyrosine), but is not thought to be an important cause of spoilage. *C. maltaromaticum* (formerly *C. piscicola*) is found in fish, and may have a role in spoilage of MAP products where again it may produce tyramine.

Corynebacterium

Corynebacteria are Gram-positive pleomorphic, non-sporeforming rods. Some species may be slightly curved or have no club ends. They are facultative anaerobic and catalase-positive. Some are pathogens, but those found in foods are generally non-pathogenic and associated with plants. Most are mesotrophs but a few strains are psychrotrophic. *Corynebacterium* spp have been isolated from raw grated beetroot and are associated with spoilage of vegetables and also fish harvested from warm waters. They may be part of the initial microflora of packaged meat products, but do not seem to be important in spoilage. They are prevalent in poultry slaughtering premises where they cross-contaminate equipment and surfaces from the skin follicles.

Erwinia

Proteolyic *Erwinia* spp are members of the Enterobacteriaceae, but are mainly associated with plants. Species such as *E. herbicola* and *E. carotovora* (now called *Pectobacterium carotovora*) are important in the decay and spoilage of vegetable products, particularly 'leafy' products, such as fresh pre-packed salads. *Erwinia* do not grow, and induce 'soft rot' of fresh produce stored at refrigeration temperatures.

Vibrio

Vibrio spp are Gram-negative, straight or curved rods, and several species are pathogens, such as *V. cholerae*, *V. parahaemolyticus* and *V. vulnificus*. However, other species are common in aquatic environments and are often psychrotrophic and halophilic. They are important spoilage organisms in seafood, but *V. costicola* are also associated with putrefactive spoilage of cured meat products.

Enterobacteriaceae

The family Enterobacteriaceae is a large group of Gram-negative, oxidase-negative, non-sporeforming, motile and non-motile, facultatively anaerobic bacteria that ferment glucose and produce acid and gas. They include several important human pathogens like *Salmonella*, *Shigella*, *E. coli* and *Yersinia*, but some are involved in the microbiological spoilage of certain products. Strains of some species, such as *Citrobacter freundii*, *Klebsiella* spp, *Providencia rettgeri* and *Enterobacter agglomerans* (some strains now classified as *Pantoea agglomerans*) have been associated with spoilage of foods, particularly vacuum-packed and MAP meat and poultry products, and cured meats. Other members of the Enterobacteriaceae include *Edwardsiella*, *Erwinia*, *Escherichia*, *Hafnia*, *Proteus* and *Serratia*.

Enterobacteriaceae are often used as an indicator of process failure, or post-process contamination due to poor hygienic practices.

Coliform Bacteria

Coliform bacteria are members of the Enterobacteriaceae. They are defined as Gram negative, rod-shaped organisms which ferment lactose to produce acid and gas at 37 °C. Coliforms are commonly used as indicators of poor sanitary practices, inadequate processing or post-processing contamination e.g. Edwardsiella, Klebsiella, or Arizona. Coliforms can be found in the aquatic environment, in soil, on vegetation and also in the faeces of warm-blooded animals.

Faecal coliforms include genera that originate from the faeces. They produce acid and gas from lactose between 44 °C and 46 °C. *E. coli* is a typical faecal coliform, and is often used as an index of faecal contamination i.e. they can indicate the presence of other enteric pathogens such as *Salmonella*. They can also be used similarly to coliforms as indicators of poor sanitary practices, inadequate processing or post-processing contamination.

Bibliography

McClure P.J. Spore-forming Bacteria, in *Food Spoilage Micro-organisms*. Blackburn C. de W. Cambridge. Woodhead Publishing Ltd, 2006.

Betts G. Other spoilage bacteria, in *Food Spoilage Microorganisms*. Eds Blackburn C. de W. Cambridge. Woodhead Publishing Ltd, 2006, 668-94.

Mc Clure P.J. Spore-forming bacteria in *Food spoilage microorganisms*. Blackburn C. de W. Woodhead Publishing Ltd, Cambridge. 2006, 579-623.

Franz C.M.A.P., Holzapfel W.H. Enterococci, in *Emerging Foodborne Pathogens*. Eds Motarjemi Y., Adams M. Cambridge. Woodhead Publishing Ltd, 2006, 557-96.

Jay J.M., Loessner M.J., Golden D.A. Modern Food Microbiology. USA. Springer, 2005.

Jay J., Loessner M., Golden D. Indicators of microbial Quality and Safety, in *Modern Foods Microbiology*. Eds Jay J., Loessner M., Golden D. New York. Springer, 2005, 473- 491.

Odhav B. Bacterial contaminants and mycotoxins in beer and control strategies in *Reviews in food and nutrition toxicity, volume 2*. Eds Preedy V.R., Watson R.R. Boca Raton. CRC Press, 2005, 1-18.

Osborne J.P., Edwards C.G. Bacteria important during winemaking in *Advances in food and nutrition research, volume 50*. Ed. Taylor S.L. London. Elsevier Academic Press, 2005, 140-77.

Terano H., Takahashi K., Sakakibara Y. Characterization of spore germination of a thermoacidophilic spore-forming bacterium, *Alicyclobacillus acidoterrestris*. *Bioscience, Biotechnology, and Biochemistry*, 2005, (June), 69 (6), 1217-20.

Walker M. Juice drink spoilage - *Alicyclobacillus acidoterrestris*. *Soft Drinks International*, 2005 (February), 26-27.

Jay J., Loessner M., Golden D. Modern Food Microbiology. New York. Springer, 2005.

Montville T.J., Matthews K.R. Food Microbiology an Introduction. Washington DC. ASM Press, 2005.

Chang S.-S., Kang D.-H. *Alicyclobacillus* spp. in the fruit juice industry: history, characteristics, and current isolation/detection procedures. *Critical Reviews in Microbiology*, 2004, 30, 55-74.

Franz C.M.A.P., Holzapfel W.H. The genus Enterococcus: biotechnological and safety issues, in *Lactic Acid Bacteria: Microbiological and Functional Aspects*. Eds Salminen S., von Wright A., Ouwehand A. New York. Marcel Dekker, 2004, 199-247.

Krovacek K., Faris A. *Aeromonas* Species, in *International Handbook of Foodborne Pathogens*. Eds Miliotis M.D., Bier J.W. New York. Marcel Dekker Inc., 2003, 357-68.

Australian Institute of Food Science and Technology Incorporated Food Microbiology Group, Moir C.J. Spoilage of processed foods: causes and diagnosis. Waterloo DC. AIFST Inc., 2001, 428.

Brackett R.E., Frank J.F., Jackson T.C., Marshall D.L., Acuff G.R., Dickson J.S. Microbial spoilage of foods, in *Food Microbiology: Fundamentals and Frontiers*. Eds Doyle M.P., Beuchat L.R., Montville T.J. 2nd edition. Washington DC. ASM Press, 2001, 91-138.

Russell S.M. Spoilage bacteria associated with poultry, in *Poultry Meat Processing*. Ed. Sams A.R. Boca Raton. CRC Press, 2000, 159-79.

Jensen N. *Alicyclobacillus* – a new challenge for the food industry. *Food Australia*, 1999, 51 (1-2), 33-6.

Hanlin J.H. Spoilage of acidic products by *Bacillus* species. *Dairy, Food and Environmental Sanitation*, 1998, 18 (10), 655-9.

Morton R.D. Spoilage of acidic products by butyric acid anaerobes – a review. *Dairy, Food and Environmental Sanitation*, 1998, 18 (9), 580-4.

Forsythe S.J., Hayes P.R. Food spoilage (I) & (II), in Food Hygiene, Microbiology and HACCP. 3rd Edn. Eds Forsythe S.J., Hayes P.R. Gaithersberg. Aspen Publishers, 1998 86-149.

International Commission on Microbiological Specifications for Foods. Microorganisms in Foods, Volume 6: Microbial Ecology of Food Commodities. London. Blackie Academic & Professional, 1998.

Commission of the European Communities, Hinton M.H., Mead G.C., Rowlings C. Microbial Control in the Meat Industry, Volume 4: Meat Spoilage and its Control. Bristol. University of Bristol Press, 1997.

Zeuthen P., Mead G.C. Microbial spoilage of packaged meat and poultry, in Meat Quality and Meat Packaging. Ed. Taylor S.A. Utrecht. ECCEAMST, 1996, 273-83.

Gram L., Huss H.H. Microbiological spoilage of fish and fish products. *International Journal of Food Microbiology,* 1996, 33 (1), 121-37.

Borch E., Kant-Muermans M.-L., Blixt Y. Bacterial spoilage of meat and cured meat products. *International Journal of Food Microbiology,* 1996, 33 (1), 103-20.

Huis in't Veld J.H.J. Microbial and biochemical spoilage of foods: an overview. *International Journal of Food Microbiology,* 1996, 33 (1), 1-18.

Brackett R.E. Microbiological spoilage and pathogens in minimally processed refrigerated fruits and vegetables, in Minimally Processed Refrigerated Fruits and Vegetables. Ed. Wiley R.C. London. Chapman and Hall, 1994, 269-312.

Gibbs P.A., Patel M., Stannard C.J. Microbial ecology and spoilage of chilled foods: a review. *Leatherhead Food Research Association*, December 1982.

Speck R.V. Thermophilic organisms in food spoilage: sulphide spoilage anaerobes. *Journal of Food Protection,* 1981, 44 (2), 149-53.

Ashton D.H. Thermophilic organisms involved in food spoilage: thermophilic anaerobes not producing hydrogen sulphide. *Journal of Food Protection,* 1981, 44 (2), 146-8.

Thompson P.J. Thermophilic organisms involved in food spoilage: aciduric flat-sourspore forming aerobes. *Journal of Food Protection,* 1981, 44 (2), 154-6.

Ito K.A. Thermophilic organisms in food spoilage: flat-sour aerobes. *Journal of Food Protection,* 1981, 44 (2), 157-63.

FOOD-SPOILAGE FUNGI

MOULDS AND MYCOTOXINS IN FOOD

Mycotoxin contamination of crops has been a worldwide problem for thousands of years. However, the significance of the mycotoxins present in foods, its effect on human health and its impact on the economy has been assessed only over the last few decades. Plant fungal pathogens or food spoilage moulds are the source of this type of toxin (1). The Food and Agriculture Organisation (FAO) estimates that 25% of the world's food crops are affected by mycotoxins during growth and storage (2). Various different types of health problems are linked to the ingestion of different types of mycotoxins by humans or animals (1). Many outbreaks related to food contamination by mycotoxins have occurred all over the world, which is a global concern as a health hazard. Summaries of the key characteristics of the main moulds and mycotoxins considered in this chapter can be found in the following tables.

Mycotoxin	Major producing fungi	Affected crops	Illness
Aflatoxins: B_1, B_2, G_1, G_2	*A. flavus*, *A. parasiticus* & *A. nomius*	Cereal grains, peanuts, corn, cottonseed, figs, most tree nuts, milk, sorghum and walnuts	Hepatocarcinoma (acute aflatoxicosis). Acute liver damage, cirrhosis, carcinogenic, teratogenic an immunosuppressive
Ochratoxins: OTA	*A. ochraceus*, *P. verrucosum* & *A. carbonarius*	Cereals, barley, beans, coffee, feeds, maze, oats, rice, rye, wheat, cheese, meat products and dry foods (fish, fruit, nuts)	Balkan nephropathy and chronic interstitial nephropathy, testicular cancer, kidney necrosis, teratogenic and immunosuppressive
Patulin	*A. clavatus* & *P. expansum*, *P. griseofulvum*, *P. roqueforti* var *carneum*, *Byssochlamys* and *Paecilomyces*	Apple, apple juice, beans, wheat, apricots, grapes, peaches, pears, olives, cereals and low acid fruit juices	Hepatotoxicity and teratogenicity affects. Increase the cell permeability and inhibits several enzymes in the cell
Fumonisins: B_1	*F. moniliforme*, *F. proliferatum* & *F. subglutinans*	Corn, sorghum and rice	Oesophageal tumours, human neural tube defects, pulmonary oedema (pigs)
Trichothecenes: T-2 toxin	*F. graminearum*, *F. equiseti*, *F. nivale*, *F. sporotrichioides*, *F. poae*, *F. chlamydosporum*, *F. acuminatum*, *F. culmorum*	Corn, feeds and hay	Alimentary toxic aleukia; variable symptoms, leukaemia; depression of immune response and inhibition of protein synthesis
Zearalenone	*F. graminearum*, *F. equiseti*, *F. culmorum* & *F. crookwellense*	Cereals, corn, feeds, rice	Hyperoestrogenism (swine), reproductive disorders (humans/animals)

Mould	Properties	Temperature (°C)			a_w			pH		
		Min	Opt	Max	Min	Opt	Max	Min	Opt	Max
A. parasiticus & *A. flavus*	Growth	6	35	54	0.81	0.95	1	2.1	4-7	11.2
	Aflatoxin production	15	33	37	0.83	0.99	0.97	3	6	8
A. ochraceus	Growth	8	30	37	0.78	0.95-0.99	>0.99	2.2	3-8	13
	OTA production by *A. ochraceus*	12	25-30	37	0.83-0.87	0.98	>0.99	-	-	-
F. graminearum	Growth	<5	24-26	<37	0.9	-	-	2.4	-	9.5-10.2
	Deoxynivalenol production by *F. graminearum*	-	24	-	-	-	-	-	-	-
	Zearalenone production by *F. graminearum*	-	25-30	-	-	>0.98	-	-	-	-
	T-2 toxin production by some *Fusarium* species	-	15	-	-	-	-	-	-	-
P. verrucosum	Growth	0	20	35	0.79-0.83	0.95	>0.99	<2.1	6-7	>10
	OTA production by *P. verrucosum*	5-10	25	-	0.83-0.85	0.90-0.95	-	-	-	-
	P. expansum									
	Growth	-2 to -6	20-25	30	<0.86	0.99	-	-	-	-
	Patulin production by *P. expansum*	-	15-20	-	-	-	-	-	-	3.5
	Yeasts growth (generally)	~0	-	~40	0.90-0.95	-	-	3	4.5-6.5	10

MYCOTOXINS

Toxins produced by food-borne fungi are termed mycotoxins; toxins produced by Basidiomycetes (e.g. mushrooms or toadstools) are not included in the mycotoxin group (3,4). Mycotoxins are secondary metabolites produced by fungi and consist of relatively small, low weight molecules (1). They have four distinct types of toxicity - acute, chronic, mutagenic and teratogenic (5). Mycotoxins are produced only at certain stages of mould growth. Fungi produce mycotoxins under stressful conditions such as changes in temperature, moisture, aeration, or the presence of aggressive agents.

The action of mycotoxins in the human body is described as a mycotoxicosis (4). Symptoms of mycotoxicoses vary depending on the quantity of toxin consumed, the chemical structure of the toxin, the target site in the body, the host species affected, and the host status (1).

The problem of mycotoxin contamination is complex for a number of reasons – the difficulties encountered in detecting the toxin in the body, the different types produced, the fact that contamination might occur anywhere in the food chain, and that a food may contain more than one type of toxin (3). In addition, many mycotoxins are heat resistant and are not easily inactivated by cooking and sterilising procedures used in food processing (6). Some mycotoxins may act synergistically, enhancing the effect of each toxin on the host body (7) e.g. co-occurrence of fumonisins and aflatoxins may cause enhanced liver carcinogenesis in humans (4). Therefore, the presence or absence of mould growth is not an indicator of the level of toxins present (7), although the isolation of toxigenic moulds should be treated as an index of the potential presence of mycotoxins, in much the same way that coliforms or Enterobacteriaceae can be used as indicators of poor sanitary practices, inadequate processing or post-process contamination, and *E. coli* an index of faecal contamination.

Field crops are more susceptible to fungal attack if drought stressed, as a result of a lack of water, overcrowding, invasion by weeds, poor soil, insect attack, or because some varieties are more susceptible to invasion by the fungus (8).

This chapter will cover the most common mycotoxins that are responsible for major epidemics in humans and animals in recent times.

Note: The literature on the subject of mycotoxin contamination can be confusing, as many fungal names have changed over the years, particularly within *Penicillium* and *Fusarium*; the attribution of a certain mycotoxin to a particular mould may have changed with time; take care with older references. Secondly, studies using pure cultures to grow mycotoxins in the laboratory have been fraught with problems, when contaminated cultures may have been used inadvertently.

AFLATOXINS

Aflatoxins are highly potent carcinogens which were discovered in the early 1960s (4,9,10,11). Aflatoxins are produced by *Aspergillus flavus*, *A. parasiticus* and *A. nomius*. Four major aflatoxins are known: B_1, B_2, G_1 and G_2. B and G indicate the blue and green fluorescent colours produced under u.v. light, and 1 and 2 indicate the major and minor compounds respectively (4). Aflatoxins M_1 and M_2 have lower toxicity than B_1 and B_2 and are formed when B_1 and B_2 are ingested by lactating cows (a proportion is hydroxylated in the liver and excreted in the cow's milk) (3).

Sources

Aflatoxins B_1, B_2, G_1 and G_2 are commonly found in corn, peanuts, most tree nuts, dried fruits, figs, milk, oil-bearing seeds or spices and certain cereals (4,6).

Illness Caused

Short term acute symptoms are due to the ingestion of large quantities of toxins, whereas long term chronic symptoms occur due to the ingestion of small quantities of toxin over a longer period. Short term symptoms of aflatoxin intoxication include ataxia, tremor, elevated temperature, anorexia, loss of appetite, weight loss, haemorrhage, bloody faeces and brown urine.

Long term effects include acute liver damage, liver cirrhosis, tumour indication and teratogenesis (3). Typical symptoms of aflatoxicoses include proliferation of the bile duct, necrosis, fatty infiltration of the liver, and hepatomas with generalised hepatic lesions (4).

OCHRATOXIN A (OTA)

Ochratoxin A is produced mainly by *Aspergillus ochraceus*, *A. carbonarius* and *Penicillium* (mainly *P. verrucosum*) (8). It is a potent nephrotoxin and teratogen and is often found in association with citrinin, penicillic acid and other mycotoxins (7,12).

Sources

Ochratoxin A is found in cereals, coffee, cocoa, dried fruit, spices, dried and smoked fish, soybeans, garbanzo beans, nuts and sometimes in grapes (6,13). Another source of contamination is the consumption of food of animal origin that has been fed with ochratoxin A contaminated feeds, e.g. pork (4).

Illness Caused

Due to its solubility in fat and slow rate of excretion from the host body, ochratoxin A accumulates in the fatty tissues (5). In humans, it is believed that ochratoxin A is carcinogenic (12) and might cause liver necrosis and kidney damage. It has been known to be fatal. High levels of consumption of toxin and a long exposure period can induce renal tumours and liver cancer (4). In test animals it is known to act as an immunosuppressor and a teratogen and can trigger mutagenicity in human cytochrome enzymes.

PATULIN

Patulin was first isolated in the 1940s from *Penicillium patulum* (1). Patulin is produced by some species of *Penicillium* mostly *P. expansum*,

P. roqueforti var *carneum* (also known as *P. carneum*) and *P. griseofulvum* (formerly known as *P. patulum*) (12), and also *Aspergillus* and *Byssochlamys* species (8).

Sources

Patulin is usually found in apples and apple-derived products, apricots, grapes, peaches, pears, olives, cereals and low acid fruit juices (12,13).

Illness Caused

Patulin is hepatotoxic and teratogenic (4), and is very toxic to both prokaryotes and eukaryotes. It is believed to affect the cell membrane by increasing its permeability and can inhibit several enzymes in the cell *in vitro* (8). Human toxicity has not been confirmed (12), although based upon adverse effects due to patulin in animal studies, the FDA believes that humans may be at risk at some level of exposure to patulin (14).

FUMONISINS

Fumonisins were first described and characterised in the late 1980s (12). They are highly water-soluble and are mainly produced by maize pathogens *Fusarium moniliforme*, *F. proliferatum*, and *F. subglutinans* (1,13). Common fumonisin mycotoxins include: FB_1, FB_2, FB_3 (7) and FB_4.

Sources

Fumonsins are found mainly in corn (6).

Illness Caused

Fumonisins have a neurotoxic effect (including uncoordinated movements and blindness) (4), and can cause liver cancer in rats. Fumonisin B_1 can alter cardiovascular and hepatic function and elevate serum cholesterol. It can also affect immune function, cause kidney and liver damage, and can result in death (7). An epidemiological link between fumonisins and human neural tube defects has been reported (15). Fumonisin B_1 is classified as a class II carcinogen because of the possibility of it being a potent cancer trigger and promoter (4). It is believed that fumonisin B_1 is related to the occurrence of oesophageal cancer in China and South Africa (1).

TRICHOTHECENES

Trichothecenes are highly toxic metabolites produced by many species of *Fusarium* (3,8), and include more than 200 different identified chemical structures (6,12). They can be divided into Type A and Type B trichothecenes, with Type A being more toxic than Type B.

Type A trichothecenes include T-2, HT-2 and diacetoxyscirpenol. Type B trichothecenes include deoxynivalenol, nivalenol and 3- and 15-acetyldeoxynivalenol.

T-2 toxin and its derivative toxin

T-2 toxin was isolated in the mid-1960s (4) and is considered to be more toxic than HT-2 toxin. Major producers are *F. acuminatum*, *F. chlamydosporum*, *F. equiseti*, *F poae* and *F. sporotrichioides* (12,16).

Diacetoxyscirpenol and monoacetylated derivatives

These are produced by *F. acuminatum*, *F. equiseti*, *F poae* and *F. sporotrichioides* (12,16).

Deoxynivalenol and its acetylated derivatives

These important mycotoxins are produced mostly by *F. graminearum* and *F. culmorum* (12).

Nivalenol and fusarenon-X

They are very similar to deoxynivalenol and its acetylated derivatives but are more toxic. Nivalenol is produced mostly by *F. poae*, *F. graminearum* and *F. nivale* whereas fusarenon-X is produced by *F. graminearum* (12).

Sources

Deoxynivalenol is found mainly in wheat and barley, and sometimes in corn (17). Nivalenol is found in barley, wheat, wheat flour and rice (17). T-2 is found in corn (in field), corn products and feed (17). Diacetoxyscirpenol is found in grains, barley, corn, rye, safflower seeds, wheat and mixed feeds (1).

Illness Caused

Trichothecenes have many different toxic effects in humans and animals due to their structural diversity. Trichothecene mycotoxicoses are difficult to distinguish due to the multi-organ affect (gastrointestinal tract, hematopoietsis, nervous, immune, hepatobiliary and cardiovascular systems). Most of the major trichothecenes are cytotoxic and cause haemorrhage, oedema, and necrosis of skin tissues (4). Trichothecene toxicity also includes inhibition of the host protein synthesis (1). Symptoms of trichothecene poisoning are vomiting, diarrhoea, anorexia, nausea, abdominal pain, dizziness, headache, and gastrointestinal inflammation as rapid responses (18). Less immediate effects of trichothecenes include leucopenia, ataxia, haemorrhaging of muscular tissue and degeneration of nerve cells (3). Trichothecenes are known as feed refusal toxins due to loss of appetite being the first observed symptom (7).

T-2 toxin can cause alimentary toxic aleukia (ATA) which is considered as the most important form of human food poisoning due to the ingestion of mouldy grains. Symptoms of ATA include fever, haemorrhagic rash, bleeding from the nose, throat and gums, sepsis, and exhaustion of the bone marrow (3). T-2 toxin causes cellular damage in the bone marrow, intestines, spleen and lymph nodes, and is considered the most potent immunosuppressant of the mycotoxins (4).

Deoxynivalenol or vomitoxin can cause feed refusal 'anorexia' and emesis in humans/animals (4). Deoxynivalenol causes nausea, vomiting and diarrhoea when consumed in high doses, and causes weight loss and food refusal when consumed in low doses (1).

ZEARALENONE

Zearalenone, also called F-2, is mainly produced by *Fusarium* genera including *F. graminearum*, *F. culmorum* and *F. equiseti* (4,8,16). It can be produced in the field or during commodity storage (18).

Sources

Zearalenone is found in cereals, wheat, corn semolina and flour (6).

Illness Caused

Zearalenone is a phytoestrogen responsible for reproductive disorders in human and animals caused by its oestrogenic effect at high concentrations (7). This is due to the ability of zearalenone to mimic the body's production of oestrogen, causing feminisation in males, and interfering with the conception, ovulation and foetal development in females (17).

FOODBORNE MOULD PRODUCING TOXINS

Fungi can grow on crops in the field as well as in stored grain, resulting in colour, textural and odour changes. *Aspergillus*, *Penicillium* and *Fusarium* species are amongst the most common fungi associated with growth in, and damage to, food crops in the field, and in store if poor storage conditions prevail after harvest (3). There is some variability in the growth data reported for many of the organisms included here.

ASPERGILLUS

The genus *Aspergillus* has been well-known for a long time, and its role in food spoilage is well-established. Mycotoxins produced by *A. flavus* were identified as the cause of 'turkey X disease' in the 1960s, when 100,000 turkey poults died in the UK.

A large number of *Aspergillus* species have been associated with food spoilage. These include *A. flavus*, *A. parasiticus*, *A. ochraceus*, *A. candidus*, *A. restrictus*, *A. niger* and *A. carbonarius* (12).

Sources

A. flavus and *A. parasiticus* are widely distributed in nature. They are mostly found in nuts and seed oils, and more specifically, peanuts, corn and cotton-seed. *A. ochraceus* is found in drying or decaying vegetation, seeds, nuts (including peanuts, pecans and betel nuts) and fruits (3).

Growth Requirements for Optimum Toxin Production

A. flavus and *A. parasiticus* are similar in their requirements. They are xerophilic. *A. ochraceus* is similar. *A. nomius* is closely related to *A. flavus*, except that it does not grow at low water activities (16).

Temperature

A. flavus grows in the temperature range of 10-43 °C (3) but optimal growth occurs at 35 °C (19). Aflatoxins are produced over a temperature range of 15-37 °C (3), with maximum production at 33 °C (19). Overall patterns of growth in *A. parasiticus* are similar to *A. flavus* (3).

Growth of *A. ochraceus* occurs in the temperature range 8-37 °C, with optimum growth at 30 °C (19). Production of ochratoxins by *A. ochraceus* occurs in the temperature range of 12-37 °C, with optimum production occurring at 25-30 °C (19). *A. ochraceus* also produces penicillic acid in the temperature range of 10-31 °C (optimum at 16 °C) (3).

Studies have shown that growth of *A. carbonarius* occurs at 32-35 °C for strains from a range of southern European countries, with no growth at <15->45 °C (19).

Water activity

Minimum a_w for growth of *A. flavus* and *A. parasiticus* is 0.81 (20). Aflatoxin production is maximised at a high a_w and relatively low temperature environment, with a minimum a_w for aflatoxin production being about 0.83 (4).

Optimum growth of *A. ochraceus* occurs at a_w 0.95-0.99 (3), although it can grow in the range of a_w 0.78->0.99 (3,20). Optimum production of OTA by *A. ochraceus* occurs at a_w of 0.98 with a minimum of 0.83-0.87 (19) and maximum of >0.99.

The minimum a_w for growth of *A. nomius* is 0.83 at 25-30 °C, and 0.81 at 37 °C (16). The optimum a_w for growth of *A. carbonarius* varies between 0.93-0.987 (19).

pH

pH has little effect on the growth of *A. flavus*. Growth can occur over a wide pH range (2.1–11.2), although growth is slower at pH <3.5. Optimum aflatoxin production occurs at pH 6, with a 50% reduction at pH 4, and very little production at pH 3. Optimal growth for *A. parasiticus* occurs between pH 4 and 7. *A. parasiticus* fails to grow at pH 2.2. Optimal growth for *A. ochraceus* occurs between pH 3-8 with a minimum pH for growth of 2.2 and a maximum pH of 13 (3).

Preservatives

Sodium benzoate, even at a low concentration, can reduce the heat resistance of *A. flavus* especially at low pH. With *A. parasiticus*, sorbic acid is necessary at a level of 1000-1500 mg/kg to prevent growth at pH 5.0-5.5; acetic acid inhibits growth at a level of 8000 mg/kg at pH 4.5. Pimaricin (natamycin) inhibits growth of *A. ochraceus* at a concentration of 20 mg/kg. Sorbic acid and pimaricin both reduce the production of OTA and penicillic acid (3).

Types of Toxins Produced

The most important mycotoxins produced by *A. flavus* and *A. parasiticus* are aflatoxins B_1, B_2, G_1 and G_2. *A. ochraceus* produces three toxins - ochratoxin A, B and C (3), with ochratoxin A being the most common. Other important toxins produced by *Aspergillus* species include cyclopiazonic acid, sterigmatocystin and patulin.

FUSARIUM

Fusarium species can cause a variety of disorders, dependant upon the organism, and the toxin it can produce. Examples of species are *F. equiseti*, *F. graminearum*, *F. moniliforme*, *F. subglutinans*, *F. proliferatum*, *F. poae* and *F. sporotrichioides* (3).

Sources

Fusarium species are mainly plant pathogens and normally occur in association with plants and cultivated soils. *Fusarium* species are responsible for wilts, blights, root rots and cankers in legumes, coffee, pine trees, wheat, corn, carnations and grasses. Infection may occur in developing seeds, and in maturing fruits and vegetables. *F. equiseti* is a widely distributed plant pathogen and soil saprophyte. *F. graminearum* can enter the human diet through cereal consumption. *F. moniliforme* is commonly present in corn. *F. sporotrichioides* is a commonly occurring species found in cereal crops, peanuts and soy beans (3).

Growth Requirements for Optimum Toxin Production

Very little information is available about *Fusarium* concerning the optimum conditions for growth and optimum toxin production requirements (3). Toxicological synergism has been reported among *Fusarium* mycotoxins (7). See table (A).

Temperature

Optimum growth of *F. graminearum* occurs between the temperature range 24-26 °C. However, growth can occur at temperatures as low as <5 °C and as high as 37 °C. Optimum deoxynivalenol toxin and zearalenone production by *F. graminearum* occurs in temperatures of 24 °C (4), and 25-30 °C (19), respectively. Optimum growth for *F. moniliforme* occurs from 22.5-27.5 °C but can range from 2.5-37 °C. Optimum growth for *F. sporotrichioides* occurs at 22.5-27.5 °C with a range of -2 °C to 35 °C (3). Optimum T-2 toxin production by some *Fusarium* species occurs at 15 °C (4).

F. chlamydosporum has minimum, optimum and maximum temperatures for growth of 5 °C, 27 °C and 37 °C respectively. Optimum temperature for *F. equiseti* growth is at 21 °C (11), with a minimum of -3 °C and maximum, depending on the isolate, of about 35 °C. The optimum temperature for growth of *F. poae* is 22.5-27.5 °C, with the minimum near 2.5-9 °C and a maximum near 32-33 °C. *F. culmorum* has been reported to be psychrotrophic, growing from 0 °C, with an optimum at 21-25 °C and a maximum of only 31 °C (16,21).

Water activity

Growth for *F. moniliforme* occurs in a_w range 0.87->0.99. Growth for *F. sporotrichioides* ranges from a_w of 0.88->0.99 (3). Optimum zearalenone production by *F. graminearum* occurs at a_w >0.98, but growth can occur at a_w of 0.90 (19).

The minimum a_w for growth of *F. equiseti* has been reported to be 0.92. The minimum a_w for growth of *F. poae* is near 0.90 between 17-25 °C. The minimum a_w for growth of *F. culmorum* is 0.87 at 20-25 °C and pH 6.5; at pH 4.0 growth does not occur below a_w of 0.90 (16).

pH

pH has very little effect on the growth of *F. graminearum*, with only a 5% variation in radial growth rates over the range pH 5-10 at 25-30 °C. Growth has been observed in the pH range of 2.4-10.2. Optimum growth for *F. moniliforme* occurs at pH of 5.5-7.5 with a minimum pH of <2.5 and a maximum of >10.6. Optimum growth for *F. sporotrichioides* occurs at a pH of 5.5-9.5, with a minimum pH of <2.5 and a maximum of >10.6 (3).

Growth of *F. equiseti* occurs at pH 3.3 but not 2.4 and is rapid at pH 10.4 (16).

Types of Toxins Produced

Fusarium mycotoxins include type A and B trichothecenes, zearalenone, fumonisins, moniliformin and fusaric acid (7). Mycotoxins produced by *F. equiseti* include nivalenol, fusarenon-X, T-2, diacetoxyscirpenol, butenolide and zearalenone. Mycotoxins produced by *F. graminearum* include deoxynivalenol, nivalenol and zearalenone. *F. moniliforme* produces fumonisin B_1 as the major mycotoxin, and fumonisin B_2. It may also produce moniliformin, fusaric acid, fusarin C and other fusarins. *F. sporotrichioides* is a major producer of T-2 toxin; additionally, some isolates may produce butenolide, fusarenon-X, neosolaniol and nivalenol (3,16,21).

PENICILLIUM

Penicillium is a large genus containing 150 recognised species, of which 50 or more occur commonly. Many species of *Penicillium* are isolated from foods causing spoilage; in addition, some may produce bioactive compounds. Penicillin was first discovered in 1929, and since that time, *Penicillium* species have been investigated for other bioactive molecules with antibiotic properties. In turn, this led researchers to recognise citrinin and patulin as 'toxic antibiotics', later termed mycotoxins. Some toxins are produced by more than one species of *Penicillium*, and many species can produce more than one mycotoxin. For example, citrinin is produced by a number of *Penicillium* species and several Aspergilli (16). Toxicity due to *Penicillium* species varies - two groups of toxins that can be distinguished are those toxins which affect liver and kidney function, and neurotoxins (causing sustained trembling in animals) (3).

Important mycotoxins produced by *Penicillium* include ochratoxin A, patulin, citrinin and penitrem A (12). Some of the most important toxigenic species in foods are *P. expansum*, *P. citrinum*, *P. crustosum* and *P. verrucosum* (3). A much larger number of *Penicillium* species are associated with food spoilage; we will confine data to these selected organisms above. Additionally, some species are associated with food fermentations, for example, *P. roqueforti* and *P. camemberti* with mould-ripened cheese, as well as causing spoilage in some cases.

Sources

Penicillium is a common soil borne saprophyte. *P. citrinum* can be found in almost every kind of food surveyed for fungi; common sources are milled grains and flour, and whole cereals, especially rice, wheat and corn. *P. crustosum* is a very common species in foods and feeds, and causes spoilage of corn, processed meats, nuts, cheese and fruit juices. *P. verrucosum* reported in temperate zones is associated with Scandinavian barley and also isolated frequently from meat products in Germany and other European countries (3). *P. expansum* can be found in fruits, specifically apple and apple products (22,23).

Growth Requirements for Optimum Toxin Production

Penicillium species usually grow optimally at relatively low temperatures; they are ubiquitous in soil, cereal grains and other foods in temperate climates, and also in cool stores and refrigerated foods worldwide (3).

Temperature

Optimal growth for *P. expansum* occurs at 20-25 °C with a minimum range of -2 to -6 °C, and a maximum of 30 °C. Optimal patulin production occurs at 15-20 °C (22). Optimal growth for *P. verrucosum* occurs at 20 °C (3) with a minimum of 0 °C and a maximum of 31 °C. Optimal production of OTA by *P. verrucosum* occurs at 25 °C, with a minimum of 5-10 °C (19). Optimal growth for *P. citrinum* occurs at 26-30 °C with a minimum of 5-7 °C and a maximum of 37-40 °C. Optimal citrinin production occurs at 30 °C with a minimum of <15 °C and a maximum of 37 °C. Optimal growth for *P. crustosum* occurs at 25 °C with a minimum of <2 °C and a maximum of 30 °C. Optimal penitrem A production occurs at 20-26 °C with a minimum of <17 °C and a maximum of 30 °C (3).

P. roqueforti is a psychrophile that grows vigorously at refrigeration temperatures but not above 35 °C (16,21). Growth for *P. griseofulvum* ranges from 4-35 °C with the optimum near 23 °C; patulin production occurs over the range 4-31 °C at 0.99 a_w and 8-31 °C at 0.95 a_w (16).

Water activity

Optimal growth for *P. expansum* occurs at a_w of 0.99 with a minimum a_w of <0.86 (22). Optimal growth for *P. verrucosum* occurs at a_w of 0.95 (19) with a minimum of 0.79-0.83 and a maximum of >0.99 (3). Optimum production of OTA by *P. verrucosum* occurs at a_w 0.90-0.95 with a minimum a_w 0.83-0.85 (19). Optimum growth for *P. citrinum* occurs at a_w 0.98-0.99 with a minimum of 0.80-0.84 and a maximum of >0.99. Maximum growth of *P. crustosum* occurs at a_w >0.99. Optimal penitrem A production occurs at a_w of 0.995 with a minimum a_w of 0.92 and a maximum a_w of 0.999 (3).

The minimum a_w for growth of *P. roqueforti* is 0.83 (21).The minimum a_w for germination of *P. griseofulvum* is 0.81 at 23 °C and 0.83 at 16 °C or 30 °C (16) and patulin is produced by the fungi down to a_w of 0.88.

pH

Patulin production by *P. expansum*, is optimal at low pH level (less than 3.5); this may be due to patulin's instability at higher pH. Growth of *P. expansum* can occur at a relatively high pH, as they can reduce the pH level of the medium while growing until the medium reaches the right pH level for patulin production (23). Optimum growth of *P. verrucosum* occurs at pH 6-7 with a minimum pH of <2.1 and a maximum of >10. Optimum growth of *P. citrinum* occurs at pH 5.0-7.0 with a minimum pH of <2.2 and a maximum of >9.7. Optimum growth of *P. crustosum* occurs at pH 4-9 with a minimum of <2.2 and a maximum of >10 (3).

Preservatives

The minimum inhibitory concentration (MIC) of sorbic acid to prevent the growth of *P. crustosum* is 2000 mg/kg at 4 °C, and 6000 mg/kg at 25 °C. For *P. verrucosum*, maximum inhibition of OTA production occurs at an MIC of 1500 mg/kg, using sorbic acid at 20 °C. The optimum water activity to inhibit OTA production by *P. verrucosum* at 1500 mg/kg of sorbic acid level is 0.95-0.99 (3).

Types of Toxins Produced

P. expansum produces patulin, which produces neurological and gastrointestinal effects. *P. citrinum* produces citrinin, which is a significant renal toxin to monogastric domestic animals, and it can also cause kidney damage after prolonged ingestion. *P. crustosum* produces penitrem A, a potent neurotoxin, ingestion of which can lead to severe brain damage or death. It has been reported in field outbreaks involving sheep, cows, horses and dogs. *P. verrucosum* produces ochratoxin A (3).

OTHER SPOILAGE MOULDS

Alternaria

Alternaria is a plant pathogen that can produce toxins in both pre- and post-harvest commodities. *Alternaria* species produce alternariols and tenuazonic acid, which can inhibit protein synthesis, chelate metal ions and form nitrosamines (4). *Alternaria* contaminates wheat, sorghum and barley in addition to various fruits and vegetables (17). Tenuazonic acid can also be produced by *Phoma sorghina* and *Pyricularia oryzae* (4), and has been associated with onyalai, a haematological disease (12).

Claviceps

Claviceps spp. can produce ergot alkaloids on cereals. Outbreaks of ergotism were the earliest recognised human mycotoxicoses and occur occasionally. Major ergot alkaloid producer species are *C. purpurea* (17) and *C. paspali* (12). *Phoma* and *Phomopsis* can infect lupin plants and lupin seeds, and produce lupinosis toxin (phomopsin). Phomopsin is a hepatotoxin that affects sheep health after grazing on lupins. *Pithomyces* can produce sporidesmin toxin which causes facial eczema in sheep. *Monascus ruber* is used commercially to produce 'red rice'; there is some evidence that it can produce citrinin (12).

FOOD SPOILAGE YEASTS

Yeasts are essentially single-celled fungi, and are common contaminants of many foods and beverages (24). Most food spoilage yeasts either have an ascomycete sexual stage, or no sexual stage. Yeasts have an important role in many fermented foods, for example, bread, beer, wine and vinegar. Some of the same species that are used in fermentations can also be spoilage agents if they are allowed to contaminate and grow in food products. About 10 yeast species are responsible for causing the majority of food and beverage spoilage. Most yeasts involved in soft drinks spoilage are categorised as either 'opportunistic spoilage' species or 'spoilage yeast' due to poor factory hygiene (25). Davenport (26,27,28) divided yeasts that can cause spoilage in soft drink factories into four groups:

Group 1 are spoilage organisms that are well adapted to growth in soft drinks, and are able to cause spoilage with very low cell numbers (as few as one cell per container). Characteristics of Group 1 yeasts include osmotolerance, they are aggressively fermentative, preservative resistance (particularly weak organic acids), and a requirement for vitamins. *Z. bailii* is a typical example of this group, and this group is typical of Pitt and Hocking's 10 key spoilage yeasts (16).

Group 2 organisms are described as spoilage or hygiene organisms. They are able to cause spoilage of soft drinks, but only if something goes wrong during manufacturing, for example, the preservative is added at too low a level or is absent, there is ingress of oxygen, there is pasteurisation failure, or there are poor standards of hygiene. These are common contaminants in factories, but are severely restricted if good hygiene practices are followed.

Group 3 organisms are indicators of poor hygiene standards; they do not grow in soft drinks, even if present at high numbers. These are typical of the yeasts found in many factories. The higher the count, the worse the hygiene state of the factory.

Group 4 yeasts are known as aliens, those out of their normal environment. An example would be *Kluyveromyces lactis*, dairy spoilage yeast.

Although yeasts can cause food spoilage, they are not known to cause food poisoning in the strict sense of the term. Typical effects of yeast growth in foods and beverages include production of acid or acid and gas, visible growth leading to cloudiness, and changes to the product texture, flavour and colour. It is estimated that fermentative yeasts cause 5% of visible spoilage in foods and beverages, whereas most of the less obvious spoilage is caused by opportunistic spoilage yeasts, due to poor hygiene practices. Ingestion of beverages spoiled by *Saccharomyces* and *Zygosaccharomyces* yeasts can cause gastrointestinal disorders which might be due to the yeasts' metabolites (24).

ZYGOSACCHAROMYCES

Zygosaccharomyces contains the most significant spoilage yeasts species in the food and drink industry (specifically, *Z. bailii*, *Z. rouxii* and *Z. bisporus*). They are osmotolerant, resistant to weak acid preservatives, and are able to ferment hexose sugars such as glucose and fructose (25). Spoilage caused by *Zygosaccharomyces* species can cause a serious physical injury, for example, exploding glass bottle due to the gas pressure generated as a result of yeast growth (25).

Sources

Foods preferred by *Zygosaccharomyces* yeasts for colonisation tend to be acidic with high concentrations of fermentable sugars. *Z. lentus* are found in dairy products, meat products, fruit and fruit products, bread, baking products, wine, chocolate, coffee, and soy products, with breweries also being a recognised source (25). *Z. rouxii* has been found in confectionery products, jam, jellies, fruit concentrates, syrups, and oriental fermented foods such as soy sauce. *Z. mellis* is found in honey (29).

Growth requirements

Yeasts can grow at low pH, low temperature, low water activities and high sugar and salt concentration (25). The temperature range for growth is influenced by environmental factors such as presence of antimicrobial compounds, high sugar or salt concentrations and water activity in the environment (29).

Temperature

Yeasts can grow over a wide temperature range, some preferring to grow at lower temperatures, and some at higher temperatures (29). The overall range is about 0-40 °C+. *Z. lentus* can grow at low temperatures and hence it is considered a potential spoilage organism in chilled food products.

pH

Generally yeasts can grow at a pH range between 3 and 10, and have an optimum pH between 4.5 and 6.5. *Z. bailii* can grow in a minimum pH of 2.2 (29).

Water activity

The majority of food spoilage yeasts can grow in a minimum a_w of 0.90 to 0.95. *Z. rouxii* is the most osmotolerant yeast known i.e. it can grow in foods with a high sugar concentration, with a water activity as low as 0.62. Its optimum a_w value for growth is 0.95 (29).

Preservatives

Most yeasts are sensitive to heating, but the presence of ascospores increases their heat resistance. Weak organic acids, for example, acetic acid, sorbic acid, and benzoic acid, are effective preservatives. *Z bailii* is a major nuisance, particularly in beverage factories, because it is resistant to acetic acid, ethanol, sorbic acid, benzoic acid and high sugar concentrates (24).

Gaseous atmosphere and growth of moulds and yeasts

Most moulds are aerobic organisms and grow best at normal atmospheric levels of oxygen. *F. moniliforme* is able to grow in an environment with 60% CO_2 and less than 0.5% O_2 (3). *Penicillium roqueforti* is able to tolerate up to 80% CO_2 and can grow with less than 0.5% O_2 (16).

Many yeasts, particularly film-forming types, are aerobes. Fermentative yeasts are usually facultative anaerobes, and are able to grow under low oxygen tension in beverages and in the internal areas of food (29).

Control

Mycotoxins are harmful compounds that must not be present in food. Achieving a mycotoxin-free food chain is almost impossible as the toxins can enter the human food chain directly from cereals, seeds, spices, fruits, beverages and other plant materials, or indirectly from food products obtained from animals fed contaminated feeds, through residues in milk, meat, eggs and their derivatives (7).

Mycotoxins are not generally produced in processed food products, as mycotoxin production is mainly a problem of raw commodities. Thus, good agricultural practices (GAPs) and good manufacturing practices (GMPs) are the major control actions to be taken to prevent mycotoxin presence in food commodities (13). Practical mycotoxin control can be achieved in two stages - prevention of fungal growth and toxin production pre-harvest, prevention of fungal growth and toxin production post harvest, and decontamination of existing mycotoxins in foods/feeds. These strategies include plant breeding for mould and insect resistance, and harvesting and storage practices that minimise contamination and growth (7).

Pre-harvest prevention activities for mycotoxins depend on the species and the crops. Preventing or minimising mould infection can be achieved using GAPs such as appropriate fertilisation and irrigation, crop rotation, use of pesticides (to reduce insect injury to plant kernels), weed control, and appropriate planting and harvesting time (8). Breeding practices can be used for developing resistance to mould infestation; these should also include pest resistance breeding programmes (13). Harvest, post-harvest and storage prevention activities can be achieved by controlling humidity (water activity), preventing insect, rodent or bird attacks, and preventing temperature and moisture gradients, which can allow mould growth in otherwise 'safe' commodities (8). Below are some examples of control.

Aflatoxins can be controlled by eliminating the fungus from the plant before harvesting by the use of selective fungicides or systemic competition by non-toxigenic fungi. Control of aflatoxins in commodities after it has been formed generally relies on sampling and screening techniques e.g. in corn, cottonseed and figs. Aflatoxins can be detected by fluorescing under ultraviolet light. Infected peanuts are discoloured and segregation of toxic nuts is carried out using colour-sorting machines. In pistachios, segregation of toxic nuts is carried out by chemical tests. In developed countries, procedures used for sorting and cleaning-up reduce aflatoxins to low levels in foods (3).

Control of the occurrence of ochratoxin is achieved by screening for the toxin, usually in commodities where high levels of *A. ochraceus* or *P. verrucosum* have been found (3).

Control of the occurrence of patulin in fruit juices is achieved by using undamaged fruit, adopting good harvesting and post harvest practices and screening for the toxin.

Good hygiene manufacturing processes can achieve control of spoilage yeasts in foods. Permitted preservatives can be added to prevent growth of spoilage yeasts. Foodstuffs can be preserved by chilling, heating or other physical methods (24).

Bibliography

References

1. Bennett J.W., Klich M. Mycotoxins. *American Society for Microbiology, Clinical Microbiology Reviews*, 2003, 16 (3), 497-516.

2. IFST Institute of Food Science & Technology Trust Fund through its Public Affairs and Technical & Legislative Committees. Information Statement. *Mycotoxins*. Prepared by IFST Professional Food Microbiology. 2006.

3. Roberts T.A., Baird-Parker A.C., Tompkin R.B. *Micro-organisms in foods 5 Microbiological specifications of food pathogens*. London, Weinheim, New York, Tokyo, Melbourne and Madras, Blackie Academic & Professional, an imprint of Chapman & Hall. 1996.

4. Riemann H.P., Cliver D.O. *Foodborne Infections and Intoxication*. The Netherlands, London, Burlington, California, Academic Press is an imprint of Elsevier. 2006.

5. Hocking A.D., Pitt J.I. Mycotoxigenic Fungi in *Foodborne Microorganisms of Public Health Significance*. Australian Institute of food science and Technology Incorporated NSW Branch, Food Microbiology Group. Waterloo, Southwood Press Pty. Ltd, 2003, 641-74.

6. French Food Safety Agency. Risk assessment for mycotoxins in human and animal food chains, Summary report. AFSSA 'Agence Francaise de Securite Sanitaire des Aliments'. 2006.

7. Diaz D. The Mycotoxin Blue Book. Nottingham, Nottingham University Press, 2005.

8. David H.W. Pesticide, veterinary and other residues in food. North America & Cambridge, Woodhead Publishing Limited & CRC Press LLC, 2004.

9. Sargent K., Carnaghan R.B.A., Allcroft R. Toxic products in ground nuts –Chemistry and Origin. Chemistry and Industry. 1963, (41), .053-55.

10. Sargent, K., Sheridan, A., O'Kelly,J., Carnaghan, R.B.A. Toxicity associated with certain samples of groundnuts. *Nature*. 1961 (192), 1096-7.

11. Austwick P.K.C., Ayerst, G. Groundnut microflora and toxicity. *Chemistry and Industry*. 1963 (41), 55-61.

12. Hocking A.D. *et al.* Advances in Food Mycology: Proceedings of the Fifth International workshop of Food Mycology, Samso, October 2003. New York. Springer. 2006.

13. Murphy P.A., Hendrich S., Landgren C., Bryant C.M. Food mycotoxins: an update. *Journal of Food Science*, 2006, 71 (5), R51-R65

14. FDA. Patulin in apple juice, apple juice concentrates and apple juice products. U.S. Food and Drug Administration, Center for Food Safety and Applied Nutrition, Office of Plant and Dairy Foods and Beverages. 2001.

15. Marasas W.F. *et al.* Fumonsins disrupt sphingolipid metabolism, folate transport, and neural tube development in embryo culture and in vivo: a potential risk factor for human neural tube defects among populations consuming fumonisin-contaminated maize. *Journal of Nutrition*, 2004, 134, 711-16.

16. Pitt J.I., Hocking A.D. Fungi and Food Spoilage. London. Blackie Academic and Professional, 1997.

17. Grain Inspection, Packers & Stockyards Administration. Grain Fungal Diseases & Mycotoxin Reference. USDA 'United States Department of Agriculture', 2006.

18. Peraica M., Domijan A.M. Contamination of food with mycotoxins and human health. *Institute for Medical Research and Occupational Health*, 2001, 52(1), 23-35.

19. Magan N., Olsen M. Mycotoxins on food – detection and control. North America & Cambridge. Woodhead Publishing Limited & CRC Press LLC, 2004.

20. Blackburn C. de W. Food spoilage microorganisms. Cambridge. Woodhead Publishing Limited, 2006.

21. Frisvad J.C., Thrane U. Fungal species and their specific production of mycotoxins in Introduction to Food – And Airborne Fungi. Robert Samson, Ellen Hoekstra, Jens Frisvad, Ole Filtenborg. The Netherlands, Central Bureau Voor Schimmelculutres, Utrecht. 2002, 321-33.

22. Marin S. *et al.* Evaluation of growth quantification methods for modelling the growth of P. expansum in an apple-based medium. *Journal of the Science of Food and Agriculture*, 2006, 86 (10), 1468-74.

23. McCallum J.L. *et al.* Factors affecting patulin production by P. expansum. *Journal of Food Production*, 2002, 65, 1937-42.

24. Querol A., Fleet G.H. Yeasts in Food and Beverages. Germany, Springer - Verlag, 2006.

25. Boekhout T., Robert V. Yeasts in food, beneficial and detrimental aspects. Hamburg. Woodhead Publishing limited, 2003.

26. Davenport R.R. Forensic microbiology for the soft drinks business. *Soft drinks Management International*, 1996 (April), 34-5.

27. Wareing P.W., Davenport, R.R. Microbiology of soft drinks and fruit juices in Chemistry and Technology of Soft drinks and Fruit Juices. Ashurst, P.R. Oxford, Blackwell Publishing Ltd, 2005, 279-99.

28. Stratford M., Hofman P.D., Cole M.B. Fruit juices, fruit drinks, and soft drinks in The Microbiological safety and Quality of food, Volume I. Eds Lund B.M., Baird-Parker T.C., Gould G.W. Gaithersburg. Aspen Publishers Inc., 2000, 836-69.

29. Deak T., Beuchat L.R. Handbook of Food Spoilage Yeasts. Boca Raton, New York, London and Tokyo. CRC Press, 1996.

30. Moss M.O. Toxigenic Fungi and Mycotoxins in The Microbiological safety and Quality of food, Volume II. Eds Lund B.M., Baird-Parker T.C., Gould G.W. Gaithersburg. Aspen Publishers Inc., 2000, 1490-1517.

Further reading

Leatherhead Food International. Contaminants in Foodstuffs - A Review of Maximum Limits, Volume 2. Leatherhead Food International, 2007.

Leatherhead Food International. Contaminants in Foodstuffs - A Review of Maximum Limits, Volume 1. Leatherhead Food International, 2006.

Hocking A.D. *et al.* Advances in Food Mycology: Proceedings of the Fifth International workshop of Food Mycology, Samso, October 2003. New York. Springer. 2006.

Querol A, Fleet G.H. Yeasts in Food and Beverages. Germany. Springer – Verlag, 2006.

Diaz D. The Mycotoxin Blue Book. Nottingham. Nottingham University Press, 2005.

FDA Office of Regulatory Affairs Division of Field Science. ORA Lab Manual, Volume IV Orientation and Training, Section 7-Mycotoxin analysis. 2005.

Magan N., Olsen M. Mycotoxins on food – detection and control. North America & Cambridge. Woodhead Publishing Limited & CRC Press LLC, 2004.

Aziz N.H., Moussa L.A. Reduction of fungi and mycotoxins formation in seeds by gamma-radiation. *Journal of Food Safety*, 2004, 24, 109–127.

Boekhout T., Robert V. Yeasts in food, beneficial and detrimental aspects. Hamburg. Woodhead Publishing Limited, 2003.

Samson R., Hoekstra E., Frisvad J., Filtenborg O. Introduction to Food – And Airborne Fungi. The Netherlands. Central Bureau Voor Schimmelculutres, Utrecht, 2002.

Food and Agriculture Organization. Manual on the application of the HACCP system in mycotoxin prevention and control. FAO Food and Nutrition paper, 2001. 73.

Coker R.D. Mycotoxins and their control: constrains and opportunities. Natural Resources Institute. 1997, 73.

Pitt J.I., Hocking A.D. Fungi and Food Spoilage. London. Blackie Academic and Professional, 1997.

Deak T., Beuchat L.R. Handbook of Food Spoilage Yeasts. Boca Raton, New York, London & Tokyo. CRC Press, 1996.

Roberts T.A., Baird-Parker A.C., Tompkin R.B. Micro-organisms in foods 5 Microbiological specifications of food pathogens. London, Weinheim, New York, Tokyo, Melbourne & Madras, Blackie Academic and Professional, an imprint of Chapman & Hall. 1996.

Gravesen S., Frisvad J., Samson R. Microfungi. Munksgaard, 1994.

IARC. Some Naturally Occurring Substances: Food Items and Constituents, Heterocyclic Aromatic Amines and Mycotoxins, Volume 56. Monographs on the Evaluation of Carcinogenic Risks to Humans . Lyon, France. International Agency for Research on Cancer, 1993.

Lacey J. Natural occurrence of mycotoxins in growing and conserved forage crops. Mycotoxins and Animal Foods. Eds Smith J.E., Henderson R.S. London. CRC Press, 1991, 363-397.

Frisvad J.C., Samson R.A. Filamentous Fungi in Foods and Feeds: Ecology, Spoilage, and ycotoxin Production. Handbook of Applied Mycology Volume 3: Foods and Feeds. Eds Arora D.K., Mukerjii K.G., Marth E.H. New York. Marcel Dekker, 1991, 31-68.

Pitt J.I. Toxigenic Aspergillus and Fusarium species. Mycotoxin Prevention and Control in Food Grain. Eds Semple R.L, Frio A.S., Hicks P.A., Lozare J.V. Regnet/AGPP, 1991, 25-32.

Pitt J.I., Dyer S.K., Olsen M. Invasion of developing peanut plant by Aspergillus flavus. *Letters in Applied Microbiolology*, 1991, 13, 16-20.

Pitt J.I., Hocking A.D. Significance of fungi in stored products. Fungi and Mycotoxins in Stored Products. Eds Champ, B.R., ighley, E., Hocking, A.D., Pitt, J.I. Proceedings of an International Conference, Bangkok, Thailand. 1991, 23-26.

ACIAR Proceedings No.36. 1991, 16-21.

Pohland A.E., Wood G.E. Occurrence of mycotoxins in food. Mycotoxins in Food. Ed. Krogh P. London. Academic Press, 1987, 35-64.

Christensen C.M. Storage of Cereal grains and their Products. St Paul. American Asssociation of Cereal Chemist Inc., 1982.

Cole R.J., Cox H. Handbook of Toxic Fungal Metabolites. New York. Academic Press, 1981.

Coker R.D. Aflatoxins: past, present and future. *Topical Science*. 1979, 21 (3), 143-162.

Scott P.M. Penicillium toxins. Mycotoxic Fungi, Mycotoxins, Mycotoxicoses. An Encylopaedic Handbook. Volume 1-Mycotoxic Fungi and Chemistry of Mycotoxins. Eds Wylie T.D., Morehouse L.G. New York. Marcel Dekker, 1977.

Methods

Hocking A.D *et al.* Advances in Food Mycology. Germany. Springer Science + Business Media, 2006.

EC Regulation No. 401/2006 of 23 February 2006, laying down the methods of sampling and analysis for the official control of the levels of mycotoxins in foodstuffs. 2006.

Watson D.H. Pesticide, veterinary and other residues in food. North America & Cambridge. Woodhead Publishing Limited & CRC Press LLC, 2004.

CAC/RCP 51-2003. Code of practice for the prevention and reduction of mycotoxin contamination in cereals, including ennex on ochratoxin A, zearalenone, fumonisins and tricothecenes. 2003.

Trucksess M.W., Pohland A.E. Mycotoxin Protocols. New Jersey. Humana Press Inc., 2001.

Barnett J.A. *et al.* Yeasts: Characteristics and identification. Cambridge. Cambridge University press, 2000.

Deak T., Beuchat L.R. Handbook of Food Spoilage Yeasts. Boca Raton, New York, London & Tokyo. CRC Press, 1996.

Samson R., Hocking A., Pitt J., King D. Modern Methods in Food Mycology. London. Elsevier, 1992.

HACCP

Introduction

The Hazard Analysis Critical Control Point (HACCP) system is a structured, systematic approach to ensuring food safety. HACCP provides a means to identify and assess potential hazards in food production and establish preventive control procedures for those hazards. A critical control point (CCP) is identified for each significant hazard, where effective control measures can be defined, applied and monitored. The emphasis on prevention of hazards reduces reliance on traditional inspection and quality control procedures, and end-product testing. A properly applied HACCP system is now internationally recognised as an effective means of ensuring food safety.

In addition, HACCP utilising all seven principles to an appropriate level is now a requirement for all food businesses in the UK and Europe, as per the new EU food hygiene legislation (see legislation chapter in this book). This ensures that small businesses are able to design a HACCP system that is appropriate for the type of business.

The HACCP concept can be applied to new or existing products and processes, and throughout the food chain from primary production to consumption. It is compatible with existing standards for quality management systems such as the ISO 9000 series, and HACCP procedures can be fully integrated into such systems.

The application of HACCP at all stages of the food supply chain is being actively encouraged, and increasingly required, worldwide. For example, the Codex Alimentarius advises that "the application of HACCP systems can aid inspection by regulatory authorities and promote international trade by increasing confidence in food safety".

Definitions

HACCP - A logic and scientifically based system, which identifies, assesses and controls hazards that are significant for food safety.

Control measure - An action or activity that can be used to prevent, eliminate, or reduce a food safety hazard to an acceptable level.

Corrective action - An action to be taken when loss of control at a CCP is indicated by monitoring.

Critical Control Point (CCP) - A step in the food chain at which control can be applied, and is essential to prevent, eliminate, or reduce a food safety hazard to an acceptable level.

Critical limit - A predetermined value for a control measure marking the division between acceptability and unacceptability.

Hazard - A biological, chemical or physical agent in, or property of, food that has the potential to adversely affect the health of the consumer.

Hazard analysis - The process of collecting and assessing information on hazards and the conditions leading to their presence, to determine which are significant for food safety and should be addressed by the HACCP system.

Monitoring - Conducting a planned sequence of observations or measurements to assess whether a CCP is under control.

Step - A raw material, location, procedure, operation or stage in the food production process.

Verification - The application of supplementary information, in addition to monitoring, to determine the effectiveness of the HACCP system.

Application of the HACCP System

The HACCP system consists of the following seven basic principles:

1. Conduct a hazard analysis (identify hazards and control measures).

2. Identify the Critical Control Points (CCPs).
3. Establish the critical limit(s) for each CCP.
4. Establish a monitoring system to ensure control of the CCP.
5. Establish the corrective action to be taken when monitoring shows that a CCP is not under control.
6. Establish verification procedures to demonstrate the effectiveness of the HACCP system.
7. Establish documentation concerning all procedures and records appropriate to these principles and their application.

It is recommended by the Codex Alimentarius that the practical application of the HACCP principles be approached according to the steps described in the logic sequence (Fig. 1). The decision tree (Fig. 2) should be used to determine the CCPs.

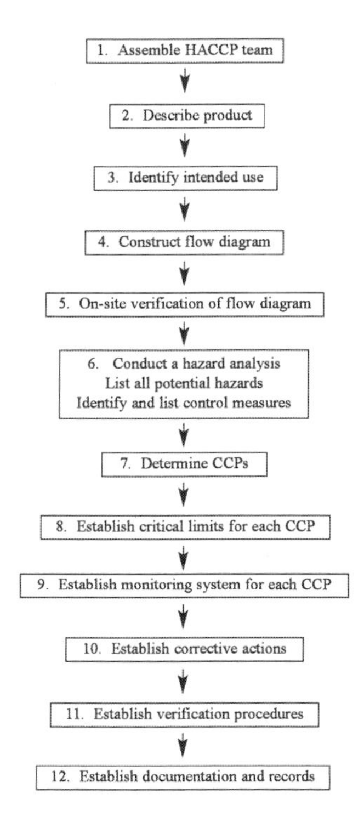

Fig. 1. Logic sequence for application of HACCP

Answer the following questions for each identified hazard:

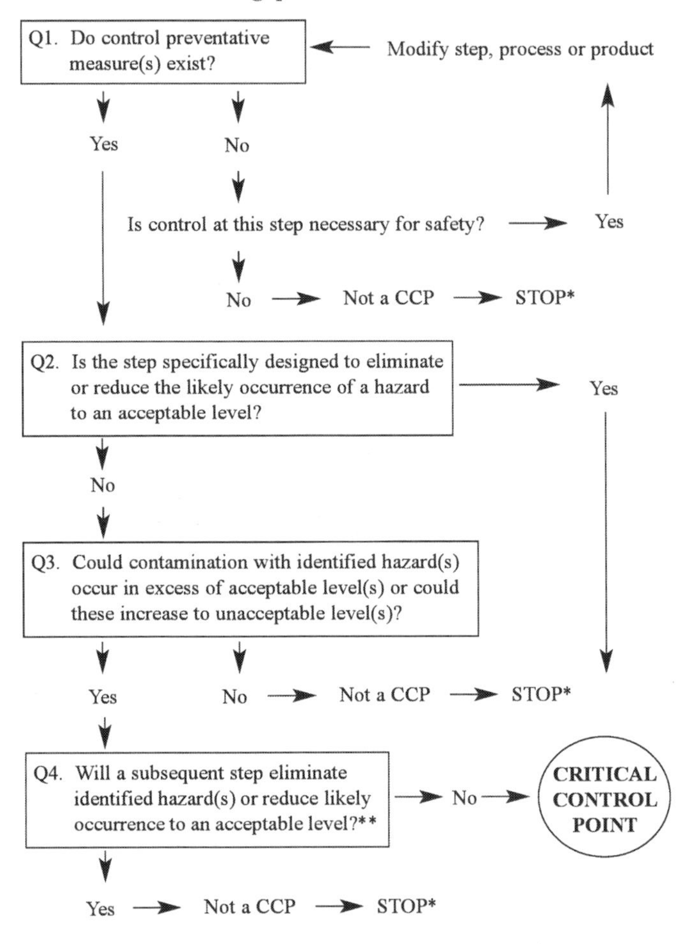

* Proceed to next step in the described process

** Acceptable and unacceptable levels need to be defined within the overall objectives in identifying the CCPs of HACCP plan

Fig.2. CCP Decision Tree A (Adapted from Codex Alimentarius Commission, 1997)

The stages of a HACCP study

1. Assemble HACCP team

An effective HACCP plan is best carried out as a multidisciplinary team exercise to ensure that the appropriate product-specific expertise is available. The team should include members familiar with all aspects of the production process as well as specialists with expertise in particular areas such as microbiology or engineering. If expert advice is not available on-site, it may be obtained from external sources.

The scope of the plan should be determined by defining the extent of the production process to be considered and the categories of hazard to be addressed (e.g. biological, chemical and/or physical).

2. Describe the product

It is important to have a complete understanding of the product, which should be described in detail. The description should include information such as composition, physical and chemical structure (including a_w, pH, etc.), processing conditions (e.g. heat treatment, freezing, smoking etc.), packaging, shelf-life, storage and distribution conditions, and instructions for use.

3. Identify intended use

The intended use should be based on the expected uses of the product by the end-user or consumer (e.g. is a cooking process required?). It is also important to identify the consumer target groups. Vulnerable groups of the population, such as children or the elderly, may need to be considered specifically.

4. Construct flow diagram

The flow diagram should be constructed by the HACCP team and should contain sufficient technical data for the study to progress. It should provide an accurate representation of all steps in the production process from raw materials to the end product. It may include details of the factory and equipment layout, ingredient specifications, features of equipment design, time/temperature data, cleaning and hygiene procedures, and storage conditions.

5. On-site confirmation of the flow diagram

The HACCP team should confirm that the flow diagram matches the process that is actually carried out. The operation should be observed at all stages and any discrepancies between flow diagram and normal practice must be recorded and the diagram amended accordingly. It is also important to include observation of production outside normal working hours, such as night shifts. It is essential that the diagram is accurate, because the hazard analysis and decisions regarding CCPs are based on these data.

6. List all potential hazards associated with each step, conduct a hazard analysis and identify control measures

The HACCP team should list all hazards that may reasonably be expected to occur at each step in the production process.

The team should then conduct a hazard analysis to identify which hazards are of such a nature that their elimination or reduction to an acceptable level is essential to the production of safe food.

The analysis is likely to include consideration of:

- the likely occurrence of hazards and the severity of their adverse health effects;
- the qualitative and/or quantitative evaluation of the presence of hazards;
- survival or multiplication of pathogenic micro-organisms;
- production or persistence of toxins.

The HACCP team should then determine what control measures exist that can be applied for each hazard.

Some hazards may require more than one control measure for adequate control and a single control measure may act to control more than one hazard.

Note: it is important at this stage that no attempt is made to identify CCPs, since this may interfere with the analysis.

7. Determine critical control points

The determination of CCPs in the HACCP system is facilitated by using a decision tree (Fig. 2) to provide a logical, structured approach to decision making. However, application of the decision tree should be flexible and its use may not always be appropriate. It is also essential that the HACCP team has access to sufficient technical data to determine the CCPs effectively.

If a realistic hazard has been identified at a step where control is required for safety, but for which no control exists at that step or any other, then the process must be modified to include a control measure.

8. Establish critical limits for each CCP

Where possible, critical limits should be specified and validated for each CCP. More than one critical limit may be defined for a single step. For example, it is usually necessary to specify both time and temperature for a thermal process. Criteria used to set critical limits must be measurable and may include physical, chemical, biological or sensory parameters.

9. Establish a monitoring system for each CCP

Monitoring involves planned measurement or observation of a CCP relative to its critical limits. Monitoring procedures must be able to detect loss of control of the CCP and should provide this information with sufficient speed to allow adjustments to be made to the control of the process before the critical limits are violated. Monitoring should either be continuous, or carried out sufficiently frequently to ensure control at the CCP. Therefore, physical and chemical on-line measurements are usually preferred to lengthy microbiological testing. However, certain rapid methods such as ATP assay by bioluminescence may be useful for assessment of adequate cleaning, which could be a critical limit for some CCPs.

Persons engaged in monitoring activities must have sufficient knowledge, training and authority to act effectively on the basis of the data collected. These data should also be properly recorded.

10. Establish corrective actions

For each CCP in the HACCP plan, there must be specified corrective actions to be applied if the CCP is not under control. If monitoring indicates a deviation from the critical limit for a CCP, action must be taken that will bring it back under control. Actions taken should include proper isolation and disposition of the affected product and all corrective actions should be properly recorded.

11. Establish verification procedures

Verification and auditing methods, procedures and tests should be used frequently to determine whether the HACCP system is effective. These may include random sampling and analysis, including microbiological testing. Although microbiological analysis is generally too slow for monitoring purposes, it can be of great value in verification, since many of the identified hazards are likely to be microbiological.

In addition, reviews of HACCP records are important for verification purposes. These should confirm that CCPs are under control and should indicate the nature of any deviations and the actions that were taken in each case. It may also be useful to review customer returns and complaints regularly.

12. Establish documentation and record keeping

Efficient and accurate record keeping is an essential element of a HACCP system. The procedures in the HACCP system should be documented.

Examples of documented procedures include:

- The hazard analysis
- Determination of CCPs
- Determination of critical limits

Examples of recorded data include:

- Results of monitoring procedures
- Deviations from critical limits and corrective actions

The degree of documentation required will depend partly on the size and complexity of the operation, but it is unlikely to be possible to demonstrate that an effective HACCP system is present without adequate documentation and records.

Implementation and Review of the HACCP Plan

The completed plan can only be implemented successfully with the full support and co-operation of management and workforce. Adequate training is essential and the responsibilities and tasks of the operating personnel at each CCP must be clearly defined.

Finally, it is essential that the HACCP plan be reviewed following any changes to the process, including changes to raw materials, processing conditions or equipment, packaging, cleaning procedures and any other factor that may have an effect on product safety. Even a small alteration to product or process may invalidate the HACCP plan and introduce potential hazards. Therefore the implications of any changes to the overall HACCP system must be fully considered and documented and adjustments made to the procedures as necessary.

Bibliography

Wareing P.W. Carnell A.C. HACCP – A Toolkit for Implementation. Leatherhead Food International, 2007.

Safer Food, Better Business. Food Standards Agency, 2005.

Scottish HACCP Working Group. Cooksafe. Food Safety Assurance System. Food Standards Agency, 2004.

Mortimore S., Mayes T. The effective implementation of HACCP systems in food processing, in *Foodborne Pathogens: Hazards, Risk Analysis and Control.* Eds Blackburn C. de W., McClure P.J. Cambridge. Woodhead Publishing Ltd, 2002, 229-56.

Bolat T. Implementation of the hazard analysis critical control point (HACCP) system in a fast food business. Food Reviews International, 2002, 18 (4), 337-71.

Bernard D. Hazard Analysis and Critical Control Point System: use in controlling microbiological hazards, in *Food Microbiology: Fundamentals and Frontiers*. Eds Doyle M.P., Beuchat L.R., Montville T.J. 2nd Edn. Washington DC. ASM Press, 2001, 833-46.

Mayes T., Mortiomore C.A. Making the most of HACCP Learning from other's experience. Woodhead Publishing, 2001.

Mortimore, S. Wallace, C.A. HACCP (Executive Briefing). Blackwell Science, 2001.

Unnevehr, L.J. The Economics of HACCP. (Costs and Benefits.). St Paul. Eagan Press, 2000

Chartered Institute of Environmental Health. HACCP in practice. London. Chadwick House Group Ltd, 2000.

Brown, M. HACCP in the meat industry. Cambridge. Woodhead Publishing Ltd, 2000.

Jouve J.-L. Good manufacturing practice, HACCP, and quality systems, in *The Microbiological Safety and Quality of Food, Volume 2.* Eds Lund B.M., Baird-Parker T.C., Gould G.W. Gaithersburg. Aspen Publishers, 2000, 1627-55.

Mortimore S., Wallace C. HACCP: a Practical Approach. 2nd Edn. Gaithersburg. Aspen Publishers, 1998.

Forsythe S.J., Hayes P.R. Food Hygiene, Microbiology and HACCP. 3rd Edn. Gaithersburg. Aspen Publishers, 1998.

Food and Agriculture Organization. Food quality and safety systems: a training manual on food hygiene and the Hazard Analysis and Critical Control Point (HACCP) system. Rome. FAO, 1998.

Codex Alimentarius Commission. Hazard Analysis Critical Control Point (HACCP) System and guidelines for its application. Food Hygiene: Basic Texts. Rome. FAO, 1997, 33-45.

National Advisory Committee on Microbiological Criteria For Foods. Hazard analysis and critical control point principles and application guidelines. 1997.

Bryan F.L., World Health Organisation. Hazard Analysis Critical Control Point Evaluations: A Guide to Identifying Hazards and Assessing Risks Associated with Food Preparation and Storage. Geneva. WHO, 1992.

EU FOOD HYGIENE LEGISLATION

General Principles

Introduction

Hygiene is an important aspect of food safety and, therefore, regulation of food hygiene plays an important part in most countries' food legislation. Good hygiene is essential to all parts of the production process, from sourcing of raw materials and structural requirements in the factory or processing facility, personnel hygiene, processing and production of food products, final product specifications (that may include microbiological criteria), transport and delivery vehicle requirements through to storage conditions at point of final sale. Over the years, microbiological standards have been used as one means of trying to ensure the microbiological quality and safety of raw materials and food products, especially where climatic conditions are warmer, and a number of such standards have been prescribed in food legislation in many countries. However, the application of such standards cannot in themselves guarantee a safe and quality product and use of microbiological standards in end-product testing alone is no substitute for good hygienic practices during production. Preventative measures within the implementation of good hygienic practices is considered more effective. With this in mind, emphasis in recent years, has been on the concepts of Good Manufacturing Practice (GMP) and Hazard Analysis Critical Control Point (HACCP) systems, within which microbiological criteria can play a part.

The key aim of European food legislation is to ensure safe and quality food products can be freely traded within the European Union, while maintaining the confidence of consumers in their purchases. Regulation (EC) 178/2002 on the General Principles and Requirements of Food Law, establishing the European Food Safety Authority and laying down protection in matters of food safety (*Official Journal of the European Communities*, **45**

(L37) 1-24), lays down food safety requirements, establishes measures needed to ensure unsafe food is not put on the market and requires systems to be in place to identify and respond to food safety problems. Food business operators have an obligation to withdraw unsafe food from the market. Ensuring food safety means all aspects of the production chain need to be taken into account; general requirements for safe food to be placed on the market are also necessary to secure the effective functioning of the internal market within the EU. It is necessary to be able to trace a food or food ingredient so a comprehensive system of traceability has been set up. The food business operator is best placed to manage the supply of his product(s), so has the primary responsibility for ensuring food safety. This Regulation provides the basis for assuring a high level of consumer protection, human health and consumer interest, while taking into account also the wide diversity of food supply in Europe.

As public health and safety is a key issue for the EU, there have been a number of Directives concerning food hygiene over the years in important areas such as the dairy and meat sectors. This piecemeal approach led to inconsistencies across sectors, and, following a complete review of the hygiene sector, revised food hygiene legislation is now in place at a European level.

Framework of current EU food hygiene legislation

The current hygiene regime, commonly referred to as "the hygiene package", covers the following key points:

- Modernisation, consolidation and simplification of previous EU food hygiene legislation;
- Application of effective and proportionate controls throughout the food chain, from primary production to sale or supply to the final consumer (the so-called 'farm to fork' approach);
- Focus on the controls needed to ensure public health;
- Clarification that the primary responsibility of food business operators is to ensure food is safely produced.

There are three key regulations in the hygiene package. These are Regulation (EC) No. 852/2004 of the European Parliament and of the

Council on the hygiene of foodstuffs (*Official Journal of the European Communities* 2004, **47** (L139) 1-54, as amended), Regulation (EC) No. 853/2004 laying down specific hygiene rules for foods of animal origin (*Official Journal of the European Communities* 2004, **47** (L139) 55-205, as amended and corrected) and Regulation (EC) No. 854/2004 laying down specific rules for the organisation of official controls on products of animal origin intended for human consumption (*Official Journal of the European Communities* 2004, **47** (L139) 206-320, as amended and corrected). The first of these concerns all food business operators; due to the nature of products of animal origin, specific additional requirements are detailed in the second of these for food businesses handling products of animal origin.

Primary producers such as farmers and growers are now included in the scope of the food hygiene legislation, in many cases for the first time. Depending on the type of business, food businesses need to be registered with the appropriate competent authority and food business operators (other than primary producers such as farmers and growers) need to have in place and maintain procedures based on Hazard Analysis Critical Control Point (HACCP) principles. The legislation has been structured so this can be applied flexibly depending on the size and nature of the food business in question.

Other implementing and transitional measures have been published, including Regulation (EC) No. 2073/2005 on microbiological criteria for foodstuffs. Although microbiological criteria have been present in different regulatory measures in legislation previously in force, this is the first time a measure has been agreed combining all of these, together with new criteria as deemed appropriate.

General hygiene rules

Regulation (EC) No. 852/2004 takes account of the necessity of establishing microbiological criteria and temperature control requirements based on risk assessment. Food business operators must ensure all stages of production, processing and distribution of their products under their control must satisfy requirements of the hygiene rules. As appropriate, food business operators must comply with microbiological criteria as specified, procedures to meet targets specified in the Regulation, comply with any temperature control regulations, maintain the cold chain and undertake sampling and analysis.

Food businesses must be clean, maintained in good condition and be protected against contamination. Suitable temperature-controlled handling and storage conditions of sufficient capacity for maintaining foods at appropriate temperatures must be available, and must be designed to allow temperatures to be monitored and recorded. Where necessary, transport must be capable of maintaining foods at appropriate temperatures and allow monitoring. The risk of contamination must be minimised. Cleaning and disinfection provisions are specified, and requirements for personal hygiene are detailed; an adequate supply of potable water must be available. No raw material may be accepted if it is known to be, or reasonably expected to be, contaminated with parasites or pathogens to such an extent that normal sorting/processing procedures would render the final product unfit for human consumption.

Food products that could support the growth of pathogens or toxin formation must not be kept at a temperature that could be a risk to health. The cold chain must not be interrupted, but limited periods outside specified temperature control are permitted to accommodate handling practicalities during preparation, transport, storage, display or service of food, provided there is no risk to health. Raw and processed materials must be separated, with sufficient refrigerated storage as required.

If foods are to be stored at chill temperatures, they must be cooled as quickly as possible from any heat processing or final preparation, to a temperature not giving any risk to health. Any thawing must minimise the risk of pathogen growth or toxin formation, and temperatures used must not be a risk to health. After thawing, foods or ingredients must be handled in an appropriate manner.

Wrapping and packing of food products must not contaminate the food; in particular, cleanliness and integrity of container construction is essential.

For products placed on the market in a hermetically sealed container, parameters such as temperature, pressure, sealing and microbiological criteria, including those monitored using automatic devices, must be checked. The process used should be an internationally recognised standard such as pasterurisation, sterilisation or UHT.

Food business operators must ensure food handlers are instructed and/or trained in food hygiene commensurate or appropriate to their food activity. Adequate training in HACCP principles must be given where necessary.

HACCP

Food business operators, other than at the level of primary production, and associated operations must, under Schedule 2 of the Regulations, put in place, implement and maintain a permanent procedure or procedures based on principles of the system of hazard analysis and critical control points (HACCP). Emphasis is placed on risk-related control, with responsibility placed on the proprietor of the food business to ensure that potential hazards are identified and systems are developed to control them. More detailed information on HACCP can be found in the previous chapter.

When any modification is made in the product, process or any step, food business operators must review the procedure and make the necessary changes to it. Food business operators must provide the competent authority with evidence of their compliance and ensure that any appropriate documents are up to date. Such documents and records must be kept for a specified period.

According to Article 4 of Regulation (EC) No 852/2004, food business operators are to comply with microbiological criteria. This should include testing against the values set for the criteria through the taking of samples, the conduct of analyses and undertaking any corrective actions needed, in accordance with food law and instructions given by the competent authority. Therefore implementing measures have been laid down concerning the analytical methods, including, where necessary, the measurement uncertainty, the sampling plan, the microbiological limits and the number of analytical units that should comply with these limits.

In some cases it has been considered appropriate to lay down implementing measures concerning the foodstuff to which the criterion applies and the points of the food chain where the criterion applies, as well as the actions to be taken when the criterion is not met. Food business operators may need to consider controls on raw materials, processing criteria, temperature and shelf-life of the product in order to ensure compliance.

Hygiene rules for products of animal origin

As products of animal origin tend to represent the highest risk to public health in food production, additional controls for these products have been introduced by means of Regulation (EC) No. 853/2004, as amended, laying down specific hygiene rules for foods of animal origin. Under this Regulation, hygiene provisions are detailed for products of animal origin in general, including registration and approval of establishments, definitions of products covered by the scope of the Regulations, identification marking and the requirement to comply with specified microbiological criteria.

In Annex III of the Regulations, more specific requirements are detailed for the following categories of products:

- meat of domestic ungulates
- poultry and lagomorphs
- wild game meat
- meat of farmed game
- minced meat, meat preparations and mechanically separated meat (MSM)
- raw milk, colostrum, dairy products and colostrum-based products
- gelatin and collagen
- frogs legs and snails
- live bivalve molluscs and other fishery products
- eggs and egg products
- animal fats and greaves

Within these requirements are temperature specifications for raw materials and products during processing, packaging and transport; the specified temperatures vary depending on the product as an example of the level of detail within the Regulation. Food business operators must initiate procedures to ensure that raw milk meets the following criteria:

(i) Raw cows' milk must meet the following standards:

Plate count 30 °C (per ml)	≤ 100,000 (1)
Somatic cell count (per ml)	≤ 400,000 (2)

(1) Rolling geometric average over a two-month period, with at least two samples per month.
(2) Rolling geometric average over a three-month period, with at least one sample per month, unless the competent authority specifies another methodology to allow for seasonal variations in levels of production.

(ii) Raw milk from other species must meet the following standard:

Plate count 30 °C (per ml)	≤ 1,500,000 (1)

(1) Rolling geometric average over a two-month period, with at least two samples per month.

(iii) However, if raw milk from species other than cows is intended for manufacture of products made with raw milk by a process that does not involve any heat treatment, food business operators must take steps to ensure the raw milk meets the following criterion:

Plate count 30 °C (per ml) ≤ 500,000 (1)

(1) Rolling geometric average over a two-month period, with at least two samples per month.

EU Regulation on microbiological criteria

Regulation (EC) No. 2073/2005 establishes microbiological criteria for a range of food products. Microbiological criteria are considered to give guidance on the acceptability of foodstuffs and their manufacturing, handling and distribution processes. The use of microbiological criteria should form an integral part of the implementation of HACCP-based procedures and other hygiene control measures. The microbiological criteria can define the acceptability of the processes, and food safety microbiological criteria set a limit, above which a foodstuff is considered unacceptably contaminated with the micro-organisms for which the criteria are set. Food business operators should decide the necessary sampling and testing frequencies, as part of their procedures based on HACCP principles and other hygiene control procedures. However, in some cases harmonised sampling frequencies have been set at Community level, particularly in order to ensure the same level of controls be performed throughout the Community.

International guidelines for microbiological criteria in respect of many foodstuffs have not yet been established. However, the Commission has followed the Codex Alimentarius guideline 'Principles for the establishment and application of microbiological criteria for foods CAC/GL 21 - 1997' and other advice in laying down microbiological criteria. Existing Codex specifications in respect of dried milk products, foods for infants and children, and the histamine criterion for certain fish and fishery products have been taken into account. The adoption of Community criteria is intended to benefit trade by providing harmonised microbiological requirements for foodstuffs and replacing national criteria, which were potential barriers to trade. The criteria set will be reviewed to take into account food safety and developments in microbiology.

An outline only of the specified criteria is given here; the full document, Regulation (EC) No. 2073/2005 (*Official Journal of the European Communities* L338, 22/12/2005, 1-26) can be accessed via www.eur-lex.europa.eu.

Annex 1 of Regulation (EC) No. 2073/2005 is divided into two chapters, one detailing food safety criteria and the other process hygiene criteria.

Food safety criteria

This chapter details the food categories for which microbiological criteria are specified, names the specific microbes and their toxins and metabolites as relevant, sampling plan (number of units comprising the sample and number of samples giving values between stated limits), limits, analytical reference method and the stage in the process chain where the stated criterion applies.

Table I overleaf illustrates this for ready–to-eat foods, with the notes below supplementing the data.

Notes to Table I:

1. n = number of units comprising the sample; c = number of sample units giving values over m or between m and M
2. For foods above m = M
3. The most recent edition of the standard shall be used
4. Regular testing against the criterion is not useful in normal circumstances for the following ready-to-eat foods:

- those that have been heat-treated or had other processing effective to eliminate *Listeria monocytogenes* when recontamination is not possible after this treatment (e.g. products heated in their final package)
- fresh, uncut and unprocessed fruit and vegetables, excluding sprouted seeds
- bread, biscuits and similar products
- bottled or packed waters, soft drinks, beer, cider, wine, spirit drinks and similar products
- sugar, honey and confectionery, including coca and chocolate products
- live bivalve molluscs

5. This criterion applies if the manufacturer is able to demonstrate, to the satisfaction of the competent authority, that the product will not exceed the limit of 100 cfu/g during shelf-life. Intermediate limits may be fixed during the process that should be low enough to guarantee that the 100 cfu/g limit is not exceeded at the end of shelf life.
6. 1 ml of inoculum is placed on a Petri dish of 140 mm diameter or on three Petri dishes of 90 mm diameter
7. This criterion applies to products before they have left the immediate control of the business operator producing them where he cannot show, to the satisfaction of the competent authority, that the product will not exceed the 100 cfu/g limit throughout shelf-life.
8. Products with pH </= 4.4 or a_w </=0.92, products with pH </= 5.0 and a_w </+ 0.94, products with shelf life less than 5 days are automatically included in this category. Other products can be included following scientific justification.

Table I
Food Safety Criteria for Ready-to-eat Food

Food category	Microorganisms/toxins, metabolites	Sampling plan[1]		Limits[2]		Analytical reference method[3]	Stage where criterion applies
		n	c	m	M		
Ready-to-eat food intended for infants and young children, ready-to-eat foods for special medical purposes[4]	*Listeria monocytogenes*	10	0	Absence in 25 g		EN/ISO 11290-1	Products placed on the market during shelf-life
Ready-to-eat foods able to support the growth of *L. monocytogenes* other than those intended for infants and for special medical purposes	*Listeria monocytogenes*	5	0	100 cfu/g[5]		EN/ISO 11290-2[6]	Products placed on the market during their shelf-life
		5	0	Absence in 25 g[7]		EN/ISO 11290-1	Before the food has left the immediate control of the food business operator who has produced it
Ready-to-eat foods unable to support the growth of *L. monocytogenes* other than those intended for infants and for special medical purposes[4,8]	*Listeria monocytogenes*	5	0	100 cfu/g		EN-ISO 11290-2[6]	Products placed on the market during their shelf life

The following products have additional food safety criteria specified by Annex 1:

Meat and meat products – minced meat and meat preparations intended to be eaten raw, minced meat and meat preparations from poultry meat intended to be eaten cooked, minced meat and meat preparations from other species than poultry intended to be eaten cooked, mechanically separated meat (MSM), meat products intended to be eaten raw except products where the manufacturing process or composition will eliminate any *Salmonella* risk, meat products from poultry intended to be eaten cooked, gelatine and collagen (*Salmonella*)

Dairy products – cheeses, butter, cream made from raw milk or milk that has undergone a lower heat treatment than pasterurisation, milk powder and whey powder, ice cream except products where the manufacturing process or composition will eliminate any *Salmonella* risk (*Salmonella*), specified cheeses, milk and whey powder that may contain staphylococcal enterotoxins (staphylococcal enterotoxins)

Egg products – egg products and ready-to-eat foods containing raw egg, except products where the manufacturing process or composition will eliminate any *Salmonella* risk (*Salmonella*)

Fish products – cooked crustaceans and molluscan shellfish (*Salmonella*), live bivalve molluscs and live echinoderms, tunicates and gastropods (*Salmonella* and *E. coli*), fishery products from fish species containing high amounts of histidine, fishery products having undergone enzyme maturation treatment in brine, manufactured from fish species containing high amounts of histidine (histamine)

Fruit and vegetable products – sprouted seeds, precut ready-to-eat vegetables, unpasteurised ready-to-eat fruit and vegetable juices (*Salmonella*)

Infant formulae and dietetic foods – specified dried infant formulae and dried dietary foods for special medical purposes intended for infants under six months of age (*Salmonella* and *Enterobacter sakazakii*)

Additional information is given in respect of interpreting the test results.

Process hygiene criteria

In Chapter 2 of Annex I, process hygiene criteria are specified for various higher risk products, with an additional requirement for action where unsatisfactory results are obtained. An example from this chapter is shown in Table II on the following page.

The following products have process hygiene criteria set:

Meat and meat products – carcases of cattle, sheep, goats and horses (aerobic colony count, Enterobacteriaceae, *Salmonella*), carcases of pigs (aerobic colony count, Enterobacteriaceae, *Salmonella*), poultry carcases of turkeys and broilers (*Salmonella*), minced meat (aerobic colony count, *E. coli*), MSM (aerobic colony count, *E. coli*), meat preparations (*E. coli*)

Milk and dairy products – pasteurised milk and other pasteurised liquid dairy products (Enterobacteriaceae), cheeses from milk or whey that has undergone heat treatment (*E. coli*), cheeses from raw milk, cheeses from milk having undergone a lower heat treatment than pasterurisation, ripened cheeses made from milk or whey having undergone pasterurisation or a stronger heat treatment, unripened soft or fresh cheeses made from milk or whey having undergone pasterurisation or a stronger heat treatment, (coagulase-positive Staphylococci), butter and cream from raw milk or milk having undergone a lower heat treatment than pasterurisation (*E. coli*), milk powder and whey powder (Enterobacteriaceae and coagulase-positive Staphylococci), ice cream and frozen dairy desserts (Enterobacteriaceae)

Infant feed - dried infant formulae and dried dietary foods for special medical purposes intended for infants below six months of age (Enterobacteriaceae)

Egg products (Enterobacteriaceae)

Fish products – shelled and shucked products of cooked crustaceans and molluscan shellfish (*E. coli*, coagulase positive Staphylococci)

Fruit and vegetable products – precut ready to eat fruit and vegetables and unpasteurised ready to eat fruit and vegetable juices (*E. coli*)

Rules for sampling and preparation of test samples are specified.

Table II
Example of Process Hygiene Criteria

Food category	Microorganisms	Sampling plan1 n	Sampling plan c	Limits[2] m M	Analytical reference method[3]	Stage where the criteria apply	Action in case of unsatisfactory results
Carcases of cattle, sheep, goats and horses	*Salmonella*	50[4]	2[5]	Absence in the area tested per carcase	EN/ISO 6579	Carcases after dressing but before chilling	Improvements in slaughter hygiene, review of process controls and of origin of animals

Notes to table:

1 n = number of units comprising the sample; c = number of sample units giving values over m or between m and M
2 For foods above m = M
3 The most recent edition of the standard shall be used
4 The 50 samples are derived from 10 consecutive sampling sessions in accordance with the sampling rules and frequencies specified in the Regulation
5 The number of samples where the presence of *Salmonella* is detected. The c value is subject to review to take account of progress in reducing *Salmonella* prevalence. Member States or regions having low *Salmonella* prevalence may use lower c values even before the review.

Application in UK law

The revised hygiene legislation, including requirements for microbiological criteria, are published at European level in the form of regulations, i.e. they are binding on the Member States from the date they come into force and there is no scope for interpretation. However, before regulations can be applied in the UK, a statutory instrument is required to enable enforcement, including detailing offences, penalties and defences. In the UK, the Food Hygiene (England) Regulations, S.I. 2006 No. 14, as amended by S.I. 2007 No. 56, and equivalent regulations in Scotland and Wales, form the appropriate legislation. In most cases, the national law does not reproduce all the technical requirements of the European legislation, but makes reference to it, citing appropriate references. There are certain areas, for example, temperature control requirements, where Member States can make their own provisions and such requirements are included in the UK regulations. General Food Safety Provisions from Resolution (EC) 178/2002 are implemented in the UK by the General Food Regulations 2004. Detailed guidance on hygiene controls is available via the Food Standards Agency Website, www.food.gov.uk.

Water

Council Directive 98/83/EC of 3 November 1998, on the quality of water intended for human consumption (*Official Journal of the European Communities* 1998, **41** (L330), 32-54), details measures to focus on compliance with essential health and quality parameters for drinking water, while allowing for the Member States to add other parameters as they see fit, if it is considered necessary to protect public health. The scope extends to water used in the food industry unless it can be shown that the use of such water does not affect the wholesomeness of the finished product. Parametric values have been based on available scientific knowledge. The precautionary principle has also been taken into account; the values have been selected to ensure that water intended for human consumption can be safely consumed on a life long basis, so giving a high level of health protection.

Member States must ensure that water intended for human consumption is wholesome and clean. For the purposes of the minimum requirements of this Directive, water intended for human consumption is wholesome and clean if it;

- is free from any micro-organisms and parasites and free from any substances that, in numbers or concentrations, constitute a potential danger to human health

- meets the minimum requirements set out in Parts A and B of Annex I of the Directive

Furthermore, in accordance with other articles and with the Treaty, Member States must take all other measures necessary to ensure water intended for human consumption complies with the requirements of this Directive. Microbiological parameters according to this Directive include *E. coli*, Enterococci, *Pseudomonas aeruginosa* and colony counts; certain parameters of microbiological concern are also included.

In England and Wales, the Water Supply (Water Quality) Regulations 2000, S.I. 2000 No. 3184, as amended by S.I. 2001 No. 2885, are applicable. The Natural Mineral Waters, Spring Water and Bottled Drinking Water Regulations 1999, S.I. 1999 No. 1540, amended by S.I. 2003 No. 666 and S.I. 2004 No. 656, revoked the 1994 Drinking Water in Containers Regulations and control bottled water that is outside the scope of natural mineral or spring water.

The quality of natural mineral and spring waters is covered by Council Directive 80/777/EEC of 15 July 1980, on the approximation of the laws of the Member States relating to the exploitation and marketing of natural mineral waters (Official Journal of the European Communities 1980, 23 (L229), 1-10), amended by Directive 96/70/EC of 28 October 1996 (Official Journal of the European Communities 1996, 39 (L299), 23/11/96, 26-8). The Directive includes details of microbiological criteria for natural mineral waters at source and during marketing. Amongst other provisions, the first amendment to the Directive, 96/70/EC, reserves the term 'spring water' for products satisfying certain microbiological and labelling requirements, as well as conditions of exploitation.

The Natural Mineral Waters, Spring Water and Bottled Drinking Water Regulations 1999, S.I. 1999 No. 1540, amended by S.I. 2003 No. 666 and S.I. 2004 No. 656 apply in UK.

Full copies of the EU regulations and directives referred to can be accessed via www.eurlex.com and copies of the UK Statutory Instruments can be accessed via www.opsi.gov.uk/stat.htm, the Office of Public Information website.

EU Legislation on mycotoxin levels in Food

Good hygiene practices will reduce the possibility of mould growth. Control of moulds as a further spoilage organism is an essential part of any hygiene plan. From a legislative perspective, controls exist to target certain mycotoxins which result from mould growth. Commission Regulation (EC) No. 1881/2006 of 19 December 2006 (Official Journal of the European Communities, 49 (L364) 5 -24) sets maximum levels for certain contaminants in foodstuffs. The Regulations' aim is to protect public health and keep contaminants at levels which are toxicologically acceptable, and it contains provisions on maximum limits for specified mycotoxins.

Maximum limits apply to the edible part of the foodstuff. For products, which are dried, diluted, processed or composed of more than one ingredient, the maximum levels applicable must correspond to those laid down by the Regulation and the following must be taken into account:

(a) changes of the concentration of the contaminant caused by drying or dilution process

(b) changes of the concentration of the contaminant caused by processing

(c) the relative proportions of the ingredients in the product

(d) the analytical limit quantification

The regulation set maximum limits for the following mycotoxins in specific food/food ingredients as indicated below:

Ochratoxin A

- dried vine fruit
- cereals
- baby foods
- coffee
- wine & grape must
- grape juice

Aflatoxins

- cereals
- nuts & dried fruit
- milk (M1)
- baby foods

- spices

Patulin

- apple products.

Fusarium toxins

- cereals (excluding rice)
- cereal flours
- bread, pastries, biscuits, cereal snacks & breakfast cereals
- pasta
- baby foods

Up to date scientific opinions regarding contaminant safety levels can be accessed through the European Food Safety Authority Website www.efsa.europa.eu.

SUPPLIERS

LABORATORY MEDIA SUPPLIERS

BioMérieux (UK) Ltd, Grafton House, Grafton Way, Basingstoke, Hants, RG22 6HY. Tel: +44 (0) 1256 461881. Fax: +44 (0) 1256 816863. www.biomerieux.com/home_en.htm.

Becton Dickinson and Company, 1 Becton Drive, Franklin Lakes, NJ 07417-1880, USA. www.bd.com.
UK sales 21 Between Towns Road, Cowley OX4 3LY. Tel: +44 (0) 1865 748844

Cherwell Laboratories Ltd, 1 Murdock Road, Bicester, Oxon, OX6 4XB. Tel: +44 (0) 1869 355500. Fax: +44 (0) 1869 355545. www.rapidmicrobiology.com/companies/88.php.

E & O Laboratories Ltd, Burnhouse, Bonnybridge, FK4 2HH, Scotland. Tel: +44 (0) 1324 840404. Fax: +44 (0) 1324 841314. www.eolabs.com.

Biotrace Ltd, 3 Lancer Court, Astmoor Industrial Estate, Runcorn, Cheshire, WA7 1PN. Tel: +44 (0) 1928 566976. Fax: +44 (0) 1928 580438. www.biotrace.co.uk

Biotrace International Plc, The Science Park, Bridgend, Wales, CF31 3NA. Tel: +44 (0) 1656 641400. Fax: +44 (0) 1656 768835 (Head Office). www.biotrace.co.uk

IDG (UK) formerly LabM, Topley House, 52 Wash Lane, Bury, BL9 6AU. Tel: +44 (0) 161 797 5729. Fax: +44 (0) 161 762 9322. www.idgplc.com.

Invitogen – formerly Life Technologies Ltd, 3 Fountains Drive, Inchinnan Drive, Paisley, PA4 9RF, Scotland. Tel: +44 (0) 141 814 6100. Fax: +44 (0) 141 814 6317. www.invitogen.com

Mast Diagnostics Ltd, Mast House, Derby Road, Bootle, Merseyside, L20 1EA. Tel: +44 (0) 151 933 7277. Fax: +44 (0) 151 944 1332. E-mail: sales@mastgrp.com. www.mastgrp.com.

Medical Wire & Equipment Co. (Bath) Ltd, Leafield Ind. Est., Potley, Corsham, Wilts, SN13 9RT. Tel: +44 (0) 1225 810 361. Fax: +44 (0) 1225 810 153.

Merck Chemicals Ltd, Boulevard Industrial Park, Padge Road, Beeston, Nottingham, NG9 2JR. Tel: +44 (0) 800 622935. Fax: +44 (0) 115 9430951. www.chemdat.info

M Tech Diagnostics, Unit-4, Station Road, Latchford,, Warrington, Cheshire, WA4 1LB. Tel: +44 (0)1925 416622. Fax: +44 (0)1925 416677 www.m-techdiagnostics.ltd.uk. E-mail queries@m-techdiagnostics.ltd.uk

Oxoid Ltd, Wade Rd, Basingstoke, RG24 8PW. Tel: +44 (0) 1256 841 144. Fax: +44 (0) 1256 463388. www.oxoid.com.

Sigma-Aldrich Co. Ltd, The Old Brickyard, New Road, Gillingham, Dorset, SP8 4XT. Tel: +44 (0) 1747 822211. Fax: +44 (0) 1747 823779. www.sigmaaldrich.com

Technical Service Consultants Ltd, The Ropewalk, Schofield St, Heywood, Lancs, OL10 1DS. Tel: +44 (0) 1706 620 600. Fax: +44 (0) 1706 620 445. www.tscswabs.co.uk.

VWR International Ltd, Hunter Boulevard, Magna Park, Lutterworth, Leicestershire, LE17 4XN. Tel: +44 (0) 1455 558600. Fax: +44 (0) 1455 558586. http://uk.vwr.com/app/Home

CULTURE COLLECTIONS

ATCC
The American Type Culture Collection (ATCC) is the World's largest and most diverse culture collection. The catalogues and further information can be obtained from:

American Type Culture Collection, 10801 University Blvd., Manassas, VA 20110-2209, USA. Tel: +1 800-638-6597. Fax: +1 703-365-2750.
E-mail for UK technical enquiries: atcc-tech@lgcpromochem.com

UKNCC
The United Kingdom National Culture Collection (UKNCC) co ordinates 11 of the UK's collection of microbial organisms. A single search can access information from all included collections. Collections included on the UKNCC are CABI (IMI), NCTC and NCYC.

Website http//www.ukncc.co.uk provides links to the various collections

CABRI
Common Access to Biological Resources and Information (CABRI) is a consortium of European collections and information centres and comprises 28 catalogues with over 100,000 items. Collections accessed from the website can be simultaneously searched.

Website http://www.cabri.org

WDCM
There are around 460 culture collections from 62 countries registered with the World Date Centre on Micro-organisms (WDCM)
Links to the collections and to further information can be found on http://wdcm.nig.ac.jp/

The WDCM is part of the World Federation for Culture Collections (WFCC)
http://wfcc.info

KIT/INSTRUMENT SUPPLIERS

Name and Address	Supplies include:
bioMérieux (see mediasuppliers) Tel: +44 (0) 1256 461881 Fax: +44 (0) 1256 816863 www.biomerieux.com	VITEK , VIDAS* and API Kits; *; BACTOMETER Mini API system; SLIDEX (*Staph.* or *Strep.*) kits; GEN PROBE: DNA Probe kits
BioControl Systems Inc. (USA) 122822 St 32nd Street Bellevue Parkway Bothell WA 98011 Tel: +1 206 487 2055 Fax: +1 206 486 2591	*Salmonella* 1 2 Test, and Assurance EIA for *Salmonella, Listeria*, and *E. coli* O157:H7. VIP tests for *E. coli* O157:H7 and *Listeria*. COLITRAK for coliforms and *E. coli* COLICOMPLETE for confirmed detection of coliforms and *E. coli*.
Biotrace Ltd The Science Park Bridgend CF31 3NA Clini-Lite, Bev-Trace™ Tel: +44 (0) 1656 641400 Fax: +44 (0) 1656 768835	Hygiene monitoring equipment (ATP system) system) Pro-tect ™, Clean-Trace™ -

Name and Address	Supplies include:
Celsis Ltd Cambridge Science Park Milton Road Cambridge CB4 4FX Tel: +44 (0) 1223 426008 Fax: +44 (0) 1223 426003	A range of systems for the rapid detection of microbial contaminants, for many industries including pharmaceutical, cosmetics, food and drink, dairy and water. Instruments include: M 1800, M 2800, M 4000, CelsisAdvance, systemSURE. Kits include: Milk microbial kit, Meat microbial kit, Fruit Juice test kit, Hygiene monitoring test kit.
Chemunex The Opas Centre St John's Innovation Park Cowley Road Cambridge CB4 OWS Tel: +44 (0) 1223 420815 Fax: +44 (0) 1223 420844	CHEMSCAN (analyser single cell detection) D COUNT (flow cytometry, automated analysers); FLUORASSURE Reagent kits (viable bacteria and yeast detection).
Cortecs Diagnostics Ltd Newtech Square Deeside Industrial Park Deeside Clwyd CH5 2NT Tel: +44 (0) 1244 288888 Fax: +44 (0) 1244 280221	HYGICULT range of hygiene monitoring (agar contact plate) kits
Becton Dickinson (see media suppliers) Tel: +44 (0) 1865 748844 Fax: +44 (0) 1865 781557	STAPH* Latex test (for *Staph. aureus* coagulase); HYCHEK* hygiene control dip slide.

Name and Address	Supplies include:
Don Whitley Scientific Ltd 14 Otley Road Shipley W. Yorks BD17 7SE Tel: +44 (0) 1274 595728 Fax: +44 (0) 1274 531197 E-mail: info@dwscientific.co.uk www.dwscientific.co.uk	WASP Spiral plater. Agents for PROTOCOT image analyser; RABIT* impedance system. Anaerobic Work Stations. (MACS) AES Media preparators. Gravimetric diluters. RAINBOW agar for VTEC. BIOLOG identification products
Dynal (UK) Ltd 10 Thursby Road Croft Business Park Bromborough Wirral Merseyside L62 5AZ Tel: +44 (0) 151 346 1234 Fax: +44 (0) 151 346 1223	Reagents for selective enrichment of *Salmonella*, *E. coli* O157 and *Listeria* based on DYNABEAD* technology Anti *Salmonella*, Anti cryptosporidium Dynabeads.
Dynex Technologies Daux Road Billingshurst Sussex RH14 9SJ Tel: +44 (0) 1403 783381 Fax: +44 (0) 1403 784397	Reagents and instrumentation for ELISAs. MICROLITE luminometers. Fluorescence reader.
Foss (UK) Ltd Parkway House Station Road Didcot Oxon, OX11 7NN Tel: +44 (0) 1235 513700 Fax: +44 (0) 1235 512718	BACTOSCAN automatic rapid microbiology system; EIAFOSS automated ELISA system for rapid pathogen screening.

Name and Address	**Supplies include:**
Hughes Whitlock Ltd Singleton Court Wonastow Road Monmouth Gwent NP5 EAH Tel: +44 (0) 1600 715632 Fax: +44 (0) 1600 715674	BIOPROBE luminometer system for detection of microbial contamination (ATP) on surfaces and in selected products.
Idexx Laboratories Ltd Milton Court Churchfield Road Chalfont St Peter Bucks SL9 9EW Tel: +44 (0) 1753 891660 Fax: +44 (0) 1753 891520 Technical Services: +44 (0) 800 581 786	Hygiene monitoring (ATP system) LIGHTENING; DWI approved rapid test for coliform and *E. coli* in water COLILERT; defined sub strate technology rapid method for basic micro biology tests SIMPLATE
International Diagnostic Kits IDG (UK) Ltd Topley House 52 Wash Lane Bury Lancashire BL9 6AU Tel: +44 (0) 161 797 5729 Fax: +44 (0) 161 762 9322	MALTHUS System V for *Salmonella, Listeria, Campylobacter* and *E. coli*; Latex *Salmonella* confirmation kit.
Life Sciences International (UK) Ltd Unit 5, The Ringway Centre Edison Road Basingstoke Hants RG21 2YH Tel: +44 (0) 1256 817282 Fax: +44 (0) 1256 817292 E-mail: 100436.3576@compuserve.com	Multiplate based MULTISKAN* for ELISA type tests; BIOSCREEN automated growth counter; LUMINOMETERS*

Name and Address	**Supplies include:**
3M Health Care Ltd 3M House Morley Street Loughborough Leicestershire LE11 1EP Tel: +44 (0) 1509 611611 Fax: +44 (0) 1509 237288 Customer Services: +44 (0) 1509 613191	PETRIFILM range of products (for TVC, coliforms, *E. coli*, including *E. coli* O157, and yeasts and moulds).
Mast Diagnostics (see Media Suppliers) Tel: +44 (0) 151 933 7277 Fax: +44 (0) 151 944 1332	Mast ID & MASTASCAN ELITE; MASTAZYME *Salmonella*; MAST ASSURE bacterial agglutination antisera (*Salmonella, Shigella, E. coli, Vibrio*). *Campylobacter* identification and biotyping systems; CRYOBANK bacterial preservation system. Full range of dehydrated culture media.
Merck Chemicals Ltd, Boulevard Industrial Park, Padge Road, Beeston, Nottingham NG9 2JR. Tel: +44 (0) 800 622935 Fax: +44 (0) 115 9430951. www.chemdat.info	Hygiene monitoring equipment (ATP system)

Name and Address	**Supplies include:**
Microgen Diagnostics Ltd 1 Admiralty Way Camberley Surrey GU15 3DT Tel: +44 (0) 1276 600081 Fax: +44 (0) 1276 600151	MICROBACT* Identification System MICROSCREEN* Rapid latex agglutination (*Salmonella, Listeria* *Staph. aureus, E. coli* O157, *Campylobacter*). rapid card tests for *Salmonella*, *E. coli* O157 and *Listeria*; biocontrol range of enzyme immunoassay and rapid card tests for *Salmonella, E. coli* O157 and *Listeria*.
Murex Biotech Central Road Temple Hill Dartford Kent DA1 5AH Tel: +44 (0) 322 277711 Fax: +44 (0) 322 273288	WELLCOLEX *Salmonella*, WELLCOLEX *Shigella* and WELLCOLEX *E. coli* for the identification of *Salmonella*, *Shigella* and *E. coli* O157. STAPHAUREX and STAPHAUREX PLUS for the identification of *Staph.* *aureus*. Agglutinating antisera for *Salmonella, Shigella, Vibrio* and *E. coli*.
Organon Teknika Ltd Science Park Milton Road Cambridge CB4 0FL Tel: +44 (0) 1223 423650 Fax: +44 (0) 1223 420264	LISTERIA TEK* kit; SALMONELLA TEK* kit; MICRO ID , incl. MICRO ID LISTERIA*; EHEC TEK.

Name and Address	Supplies include:
Oxoid Ltd (see Media suppliers) Tel: +44 (0) 1256 841144 Fax: +44 (0) 1256 463388	Latex agglutination kits for bacterial toxins (RPLA). OXOID SALMONELLA RAPID TEST (OSRT)*. LISTERIA RAPID TEST (OLRT) *Campylobacter* dry spot test. STAPH Latex Tests. *Esch. coli* O157 latex test
Palintest Ltd Palintest House Kingsway, Team Valley Trading Estate Gateshead Tyne and Wear NE11 0NS Tel: +44 (0) 191 491 0808 Fax: +44 (0) 191 482 5372 E-mail: palintest@palintest.com	COLILERT for coliforms and *E. coli*.
Prolab Diagnostics Unit 7, Westwood Court Clayhill Industrial Estate Neston South Wirral Cheshire L62 3UJ Tel: +44 (0) 151 353 1613 Fax: +44 (0) 151 353 1614	PROLEX O157 Latex test (for *E. coli* O157), PROLEX Rapid Acid Extraction Streptococcal Grouping kit (for Group D streptococci/enterococci), PROLEX Staph. latex agglutination kit (*Staph. aureus*) *Salmonella* and *Clostridium perfringens* antisera and reagents for *Legionella*.

Name and Address	Supplies include:
Rhône diagnostics Technologies Ltd West of Scotland Science Park Unit 3.06 Kelvin Campus Glasgow G20 0SP Scotland Tel: +44 (0) 141 945 2924 Fax: +44 (0) 141 945 2925 E-mail: rdt@rhone diagnostics.co.uk	*Salmonella* - diagnostic kits: screen - LOCATE, confirmation - SPECTATE; *Listeria* - screen Hygiene monitoring (ATP system)
Tecra Diagnostics UK Batley Business and Technology Centre Technology Drive Batley West Yorkshire WF17 6ER Tel: +44 (0) 1924 441255 Fax: +44 (0) 1924 441611	Immunoassay kits for *Listeria, E. coli* O157, *Salmonella*, BDE toxin and *Staph. aureus* Staphylococcal enterotoxin and *Bacillus* toxin. Also, UNIQUE *Salmonella* and *Listeria* test.
Tepnel Life Sciences Unit 8 St Georges Court Hanover Business Park Altrincham NA14 5UA Tel: +44 (0) 161 927 3400 Fax: +44 (0) 161 927 3401	DARAS instrument for automatic nucleic acid based tests for food pathogens.

ADDRESSES OF AUTHORITIES/SOURCES OF FURTHER INFORMATION, ETC.

Biscuit, Cake, Chocolate and Confectionery Alliance (BCCCA)
6 Catherine Street
London
WC2B 5JJ
Tel: +44 (0) 20 7420 7200
Fax: +44 (0) 20 7420 7201
office@bccca.org.uk
www.bccca.org.uk

British Hospitality Association (BHA)
Queens House
55 56 Lincoln's Inn Fields
London
WC2A 3BH
Tel: +44 (0) 20 7404 7744
Fax: +44 (0) 20 7404 7799
bha@bha.org.uk
www.bha.org.uk

British Retail Consortium
2nd Floor
21 Dartmouth Street
London
SW1H 9BP
Tel: +44 (0) 20 7854 8900
Fax: +44 (0) 20 7854 8901
www.brc.org.uk

British Standards Institution (BSI)
389 Chiswick High Road
London
W4 4AL
Tel: +44 (0) 20 8996 9001
Fax: +44 (0) 20 8996 7001
:cservices@bsi-global.com
www.bsi-global.com

CABi Bioscience (formerly International Mycological Institute)
Nosworthy Way
Wallingford
Oxfordshire
OX10 8DE
Tel: +44 (0) 1491 832111
Fax: +44 (0) 1491 829292
E-mail: enquiries@cabi.org
www.cabi.org

Campden and Chorleywood Food RA (CCFRA)
Station Road
Chipping Campden
Gloucestershire
GL55 6LD
Tel: +44 (0) 1386 842000
Fax: +44 (0) 1386 842100
:info@campden.co.uk
www.campden.co.uk

Central Public Health Laboratory (CPHL) and Communicable Disease Surveillance Centre (CDSC).

Health Protection Agency
Centre for Infections
61 Colindale Avenue
London
NW9 5HT (FHL)
NW9 5EQ (CDSC)
Tel: +44 (0) 20 8200 4400
Fax: +44 (0) 20 8200 8264 (FHL)
Fax: +44 (0) 20 8200 7868
(CDSC)
www.hpa.org.uk

Chartered Institute of
Environmental Health
Chadwick Court
15 Hatfields
London
SE1 8DJ
Tel: +44 (0) 20 7928 6006
Fax: +44 (0) 20 7827 5862
Email: info@cieh.org
www.cieh.org

Chilled Food Association Ltd
(CFA)
PO Box 6434
Kettering
NN15 5X7
Tel: +44 (0) 1536 514365
Fax: +44 (0) 1536 514395
E-mail: cfa@chilledfood.org
www.chilledfood.org

Codex Alimentarius Commission
(CAC)
Food & Agriculture Organization
of the United Nations (FAO)
Viale delle Terme di Caracalla
00153 Rome
Italy
Tel: +39 06 57051
Fax: +39 06 5705 4593
E-mail: codex@fao.org
www.codexalimentarius.net

Consumers Association (CA)
2 Marylebone Road
London
NW1 4DF
Tel: +44 (0) 20 7770 7000

Department of Health (DH)
Richmond House
79 Whitehall
London
SW1A 2NS
Tel: +44 (0) 20 7210 4850
Fax: +44 (0) 20 7210 5025
dhmail@dh.gsi.gov.uk
www.dh.gov.uk

European Chilled Food Federation
(ECFF)
c/o Chilled Food Association
PO Box 6434
Kettering
NN15 5XT
Tel: +44 (0) 1536 514365
Fax: +44 (0) 1536 515395
cfa@chilledfood.org
www.ecff.net

Food and Drink Federation (FDF)
6 Catherine Street
London
WC2B 5JJ
Tel: +44 (0) 20 7836 2460
Fax: +44 (0) 20 7836 0580
generalenquiries@fdf.org.uk
www.fdf.org.uk

Food Standards Agency
Aviation House
125 Kingsway
London
WC2B 6NH

FSA Information Centre : +44 (0) 20 7276 8181
FSA Helpline : +44 (0) 20 7276 8829
helpline@foodstandards.gsi.gov.uk
http://www.food.gov.uk/

Hannah Research Institute
Hannah Research Park
Mauchline Road
Ayr
KA6 5HL
Tel: +44 (0) 1292 674000
Fax: +44 (0) 1292 674004

The Stationery Office Publications Centre
PO Box 276
LONDON
SW8 5DT
Tel: +44 (0) 171-873-9090
www.opsi.gov.uk

Institute of Hospitality (formerly HCIMA)
Trinity Court
34 West Street
Sutton
Surrey
SM1 1SH
Tel: +44 (0) 20 8661 4900
Fax: +44 (0) 20 8661 4901
commdept@instituteofhospitality.org
www.hcima.org.uk

Institute of Food Research (IFR)
Norwich Research Park
Colney Lane
Norwich
NR4 7UA
Tel: +44 (0) 1603 255000
Fax: +44 (0) 1603 507723
ifr.communications@bbsrc.ac.uk
www.ifr.ac.uk

Institute of Food Science and Technology (IFST)
5 Cambridge Court
210 Shepherds Bush Road
London
W6 7NL
Tel: +44 (0) 20 7603 6316
Fax: +44 (0) 20 7602 9936
info@ifst.org
www.ifst.org

International Office of Cocoa, Chocolate and Sugar Confectionery (IOCCC)
Rue Defacqz 1
1050 BXL Brussels
Belgium
Tel: +32 2 539 1800
Fax: +32 2 539 1575

Veterinary Laboratories Agency
Woodham Lane
New Haw
Addlestone
Surrey
KT15 3NB
Tel: +44 (0) 1932 341111
Fax: +44 (0) 1932 347046
enquiries@vla.defra.gsi.gov.uk

National Consumers Council (NCC)
20 Grosvenor Gardens
London
SW1W 0DH
Tel:+44 (0) 020 7730 3469
Fax: +44 (0) 20 7730 0191
info@ncc.org.uk
www.ncc.org.uk

PHLS Food Microbiology
External Quality Assessment
Scheme
Food Hygiene Laboratory
PHLS Central Public Health
Laboratory
61 Colindale Avenue
London
SW9 5HT
Tel: +44 (0) 20 8200 4400
Fax: +44 (0) 20 8200 7874

Restaurant Association
Queens House
55 56 Lincoln's Inn Fields
London
WC2A 3BH
Tel: +44 (0) 20 7404 7744
Fax: +44 (0) 20 7404 7799.

Royal Institute of Public Health &
Hygiene
28 Portland Place
London
W1B 1DE
Tel: +44 (0) 20 7580 2731
Fax: +44 (0) 20 7580 6157
www.riph.org.uk

The Royal Society for the
Promotion of Health
38A St George's Drive
London
SW1V 4BH
Tel: +44 (0) 20 7630 0121
Fax: +44 (0) 20 7976 6847
rsph@rsph.org
www.rsph.org

Scottish Centre for Infection and
Environmental Health
Clifton House
Clifton Place
Glasgow
G3 7LN
Tel: +44 (0) 141 300 1100
Fax: +44 (0) 141 300 1170

Scottish Consumer Council
Royal Exchange House
100 Queen Street
Glasgow
G1 3DN
Tel: +44 (0) 141 226 5261
Fax: +44 (0) 141 221 0731
www.scotconsumer.org.uk

The Food Standards Agency
Scotland
6th Floor
St Magnus House
25 Guild Street
Aberdeen
AB11 6NJ
Tel: +44 (0) 1224 285100
scotland@foodstandards.gsi.gov.u
k

Society for Applied Microbiology
(formerly Society for Applied
Bacteriology)
Bedford Heights
Brickhill Drive
Bedford
MK4 7PH
Tel: +44 (0) 1234 326661
Fax: +44 (0) 1234 326678
www.sfam.org.uk

Society of Food Hygiene
Technology
The Granary
Middleton House Farm
Tamworth Road
Middleton
Staffordshire

B78 2BD
Tel: +44 (0) 1827 872500
Fax: +44 (0) 1827 875800
admin@sofht.co.uk
www.sofht.co.uk

United Kingdom Accreditation Service (UKAS)
21 47 High Street
Feltham
Middlesex
TW13 4UN
Tel: +44 (0) 20 8917 8400
Fax: +44 (0) 20 8917 8500
info@ukas.com
www.ukas.com

The Food Standards Agency Wales
11th Floor
South Gate House
Wood Street
Cardiff
CF10 1EW
Tel: +44 (0) 2920 678999
wales@foodstandards.gsi.gov.uk

World Health Organization (WHO)
Avenue Appia 20
1211 Geneva 27
Switzerland
Tel: +41 22 791 2111
Fax: +41 22 791 3111
info@who.int
www.who.int
Regional Office for Europe
Toxicology and Food Safety
8 Scherfigsvej
DK 2100 Copenhagen
Denmark
Tel: +45 39 171717
Fax: +45 39 171818
postmaster@euro.who.int
www.euro.who.int

Foodborne Infections
WHO Centre for Surveillance of Foodborne Infections and Intoxications
Institute of Veterinary Medicine
Robert von Ostertag Institute
Postfach 330013
D 14191 Berlin
Germany
Tel: +49 30 8412 0 2154/2156
Fax: +49 30 8412 4741

FAO/WHO Collaborating Centre for Research and Training in Food Hygiene & Zoonoses
Robert von Ostertag Institute
Postfach 330013
D 14191 Berlin
Germany
Tel: +49 30 7236 2156
Fax: +49 30 7236 2957

INTERNET

Introduction

A great deal of information of use to food microbiologists and those with an interest in food safety is now accessible through the Internet. The following section is intended to be a list of some of the most useful sites available at the time of writing. There has been an explosion in the number of Web sites containing food safety and microbiology information, and it would be impossible to list all the relevant sites. Those listed below represent the range of resources to be found, and many of them contain links to other sites.

The pace of change on the Internet is rapid, and site addresses and contents are often modified. Therefore it is worth visiting sites of interest on a regular basis and noting any significant changes. It is also important to realise that there are no effective controls over the information posted on Internet sites, and it is necessary to ensure that any information that you may want to use professionally comes from a reputable source. There are existing sites professing to give food safety information that are at best misleading.

IFST (INSTITUTE OF FOOD SCIENCE & TECHNOLOGY)
http://www.ifst.org

A source of detailed, topical information and with comprehensive links to other sites. Now has a search facility.

FDA CENTRE FOR FOOD SAFETY & APPLIED NUTRITION
http://vm.cfsan.fda.gov/

A very useful site that covers a wide range of information, including the 'Bad Bug Book', a series of summaries of the properties of food poisoning organisms.

Government/international organisations

FSA (Food Standards Agency)
http://www.food.gov.uk/

DEFRA (Department for Environment, Food and Rural Affairs)
http://www.defra.gov.uk

DOH (Department of Health - UK)
http://www.doh.gov.uk

HPA (Health Protection Agency)
http://www.phls.co.uk

Trading Standards
http://www.tradingstandards.gov.uk/index.cfm

USDA (United States Department of Agriculture)
http://www.usda.gov/

FDA (Food & Drug Administration - US)
http://www.fda.gov

CDC (Centers for Disease Control and Prevention)
http://www.cdc.gov/

WHO (World Health Organisation)
http://www.who.int/en

FAO (United Nations Food & Agriculture Organisation)
http://www.fao.org/

EFSA (European Food Safety Authority)
http://www.efsa.europa.eu/en.html

EUROPA
http://europa.eu/pol/food/index_en.htm

Eurosurveillance
http://www.eurosurveillance.org/index-02.asp

Food Standards Australia New Zealand
http://www.foodstandards.gov.au/

FSIS (Food Safety and Inspection Service)
http://www.fsis.usda.gov/News_&_Events/index.asp

Canadian Food Inspection Agency
http://www.inspection.gc.ca/english/toce.shtml

Institutes, associations and societies

APHA (American Public Health Association)
http://www.apha.org/

AOAC (Association of Offical Analytical Chemists)
http://www.aoac.org/

American Food Safety Institute
http://www.americanfoodsafety.com

IFR (Institute of Food Research - UK)
http://www.ifrn.bbsrc.ac.uk/

IFT (Institute of Food Technologists - US)
http://www.ift.org/

IFIC (International Food Information Council Foundation)
http://www.ific.org/index.cfm

National Institutes of Health (US)
http://www.nih.gov/

Society for General Microbiology (UK)
http://www.socgenmicrobiol.org.uk/

Food safety and microbiology

Food Safety Consortium
http://www.uark.edu/depts/fsc

Food Safety Website
http://www.ces.ncsu.edu/index.php?page=foodsafetyprocessing

International HACCP Alliance
http://haccpalliance.org

FoodHACCP.com - Food Safety Information Web Site
http://www.foodhaccp.com/indexcopy.html

Outbreak
http://www.outbreak.org

Seafood Network Information Centre
http://seafood.ucdavis.edu/

The Microbiology Network
http://microbiol.org/

USDA Food Safety & Inspection Service
http://www.usda.gov/agency/fsis/homepage.htm

Foodlink
http://www.foodlink.org.uk/aboutus.asp

Methods

AOAC
http://www.AOAC.org/testkits/microbiologykits.htm

BAM (Bacteriological Analytical Manual) Online
http://vm.cfsan.fda.gov/~ebam/bam-toc.html

FoodHaccp.com – New Control Methods for Food Safety
http://foodhaccp.com/

Health Canada - Compendium of Analytical Methods
http://www.hc-sc.gc.ca/food-aliment/mh-dm/mhe-dme/compendium/volume_3/e_index.html

RapidMicrobiology.com
http://www.rapidmicrobiology.com/index.php

Health Cananda – Offical Methods for the Microbiological Analysis of Food
http://www.hc-sc.gc.ca/fn-an/res-rech/analy-meth/microbio/volume1/index_e.html

ISO (International Organisation for Standardization)
http://www.iso.org/iso/en/CatalogueListPage.CatalogueList?ICS1=67&ICS2=&ICS3=&scopelist

INDEX